More praise for *Boundaries in an Overconnected World*

"Now that technology has made options out of so many interpersonal boundaries, finding one's comfortable place in the wired world requires both wisdom and a good game plan. Anne Katherine provides both, proving yet again that she can see clearly ahead of the curve."

— Michael Prager, professional speaker and author of *Fat Boy Thin Man*

"Finally a simple, wise, and realistic GPS for our adventures into cyberspace! Anne Katherine's extensive experience with boundary issues with people and food is now applied to electronic media. Combining practical experience with realistic healthy choices makes this book a rare and extremely effective tool for long-term self-protection, enjoyment, and effective functioning in the new electronic world. Anne's approach is so clear and practical; you will be blessed by her knowledge and expertise."

— H. Theresa Wright, MS, RD, LDN, author of *Your Personal Food Plan Guide*

"Anne Katherine is right on time with a terrific book. We live in a world that's seeking connection, meaning, and purpose, and our technology in cyberspace is destroying our ability to connect. Katherine reveals to us the entire iceberg, while most of us can only see the tip. A must-read for anyone with a mobile device — especially parents!"

— Mary Bellofatto, MA, LMHC, CEDS, NCC, TEP, trauma and addictions specialist

BOUNDARIES IN AN OVERCONNECTED WORLD

Books by Anne Katherine

Boundaries: Where You End and I Begin
Anatomy of a Food Addiction
Where to Draw the Line
When Misery Is Company
How to Make Almost Any Diet Work
Lick It! Fix Her Appetite Switch
Your Appetite Switch
Penumbra 1: Lifetimes of a Soul
Penumbra 2: A Soul's Journey

BOUNDARIES IN AN OVERCONNECTED WORLD

SETTING LIMITS TO PRESERVE YOUR FOCUS, PRIVACY, RELATIONSHIPS, AND SANITY

ANNE KATHERINE

NEW WORLD LIBRARY

New World Library
Novato, California

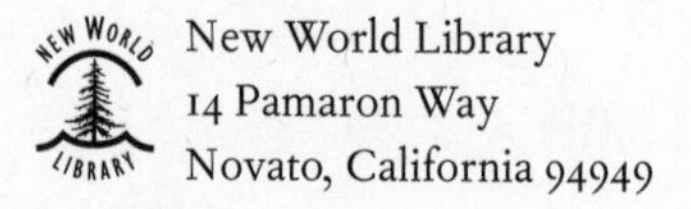
New World Library
14 Pamaron Way
Novato, California 94949

Text design by Tona Pearce Myers

Library of Congress Cataloging-in-Publication Data
Katherine, Anne.
Boundaries in an overconnected world : setting limits to preserve your focus, privacy, relationships, and sanity / Anne Katherine, MA.
pages cm
Includes bibliographical references and index.
ISBN 978-1-60868-190-7 (pbk. : alk. paper) — ISBN 978-1-60868-191-4 (ebook)
1. Interpersonal communication—Psychological aspects. 2. Information technology—Psychological aspects. 3. Interpersonal relations. 4. Privacy, Right of. 5. Distraction (Psychology) 6. Online etiquette. I. Title.
BF637.C45K358 2013
302—dc23 2013015530

First printing, September 2013
ISBN 978-1-60868-190-7
Printed in Canada on 100% postconsumer-waste recycled paper

New World Library is proud to be a Gold Certified Environmentally Responsible Publisher. Publisher certification awarded by Green Press Initiative. www.greenpressinitiative.org

10 9 8 7 6 5 4 3 2 1

For Frances West,

my chosen sister, and one of the best women I know

CONTENTS

PART IV: CYBERSPACE TRICKERY

PART V: INTIMACY

PART VI: GROUP CYBER-BOUNDARIES

Introduction

CONNECTED, BUT NOT PROTECTED

Twenty-five years ago, when I wrote *Boundaries: Where You End and I Begin*, my computer displayed green type and no pictures, and it responded to commands that were like a secret code. It was an island separate from all other computers. It couldn't talk to any other computer in the world, nor could it listen, existing prior to publicly available digital streams or networks.

The day I planned to send the manuscript to the publisher — by printing it on real paper and putting it into a mailbox — I did a spell check that, capriciously, removed the formatting from the whole manuscript. I had to go through the entire thing, page by page, and restore it manually. (Praise be for today's Undo button.)

It didn't help that I was also getting married that day, and that my "make her beautiful" appointments were threatened by this time-eating glitch. Somehow, I got it all done — including the wedding — but how different the computer world is today.

Today, total strangers can enter your home electronically, depositing repulsive ads and ridiculous offers. Friends can interrupt your focus every two minutes with texts, tweets, or adorable pictures of animals. Your personal information can land in the hands of someone you'd inch away from on a subway.

Boundaries in an Overconnected World will help you regain authority over your time, your focus, and your energy. It shows how to set limits on intrusions and exposure in cyberspace, with every communication device you own, and with everyone who wants to reach you through them.

This book doesn't replace my first book on the subject, *Boundaries*, which provides the basics of gaining autonomy in live person-to-person interactions. Nor does it replace my book *Where to Draw the Line*, which describes how to handle all sorts of situations — an indirectly hostile mother-in-law, a holiday spoiler, food pushers, sisters who take without asking, etc. — by setting specific boundaries.

Together, these three books offer a trilogy of guidance about how to set appropriate limits on earth and in space, from friends and family to strangers, and from face-to-face interactions to smartphones and laptops, so that your life is truly your own.

PART I

BOUNDARY BASICS

Chapter I

BOUNDARIES IN A BOUNDARYLESS WORLD

We now travel through an expanded universe, moving at startling speeds and negotiating vast magnitudes of miniature bits of information. We are connected not only with nearly every other computer on the planet but also with millions of minds holding a wide range of values, perspectives, needs, and goals.

The sheer number of potential connections is so great — both for information coming in and information going out — that we let our minds pretend that an email sent to twenty of our closest friends will stop in those twenty computers. But if even one recipient finds something worth passing on to Hateful Hattie — and if Hattie takes snippets of your missive out of context and sends it to her list so that her cousin Callous, a stranger to you, posts your words on a wall or in a chat room — your comments could offend someone in Singapore within minutes. You could then be Googled, found, sworn at, criticized, and targeted with offers of hemorrhoid cream and weird sexual devices.

Now that the entire electronic universe hovers within our phones, tablets (e.g., iPads), and computers, a world full of interruptions and intrusions is pinging toward us. We have a whole new set of obligations: people to (be)friend, emails to manage, texts to answer. It's easy to feel as if you'll never catch up.

If ever boundaries were needed, they are needed now, in this universe without boundaries.

This book will show you how to put virtual fences into the virtual world so that you can get your time back, and so that you can focus on what matters to you most. You will see how to free your energy for your own creative pursuits, for your beloved people, for fun and hobbies, and for sheer peace of mind.

What Is a Boundary?

A boundary is a limit that protects the integrity, autonomy, or wholeness of an entity. If your life feels out of control, unmanageable, or chaotic in any way, the right boundary will improve it.

You, an individual, are an entity. You and your best friend are also an entity — a friendship. Your family is an entity, too, and so is your unit or department at work.

It's obvious that your skin creates a physical boundary, inside of which is you and outside of which is not-you. Without that boundary you would die, too exposed to the dangers of the world and without a container for the structures that require insulation and protection.

Your best friendship also needs a boundary in regard to what comes in and what goes out. If you reveal your friend's private information (letting too much out) or keep yourself too closed off to your friend (holding too much in), the friendship's integrity will be threatened. Without a healthy boundary, a friendship can also die.

A Boundary Is a Regulator

A boundary regulates the flow of energy and information coming in and going out. Imagine that the following circle represents you.

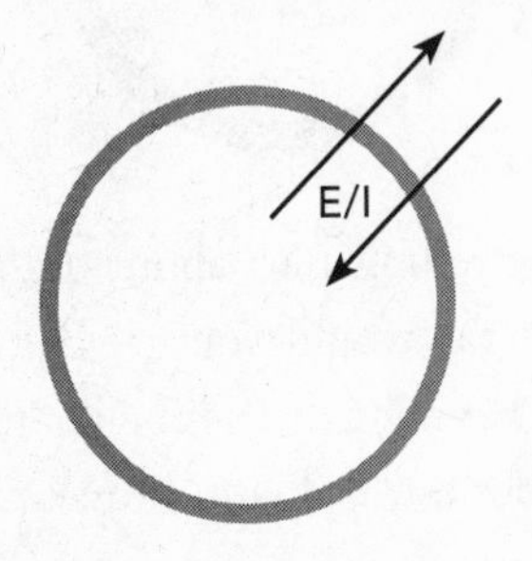

You send energy and information into the world. And you receive energy and information from outside yourself.

Have you ever met someone who is totally closed off? A pry bar can't get anything out of Blocked Bart. He gives minimal answers to questions. His face is shut. He contributes nothing to an interaction. No matter what you say, it appears to bounce off his surface.

Which of the following circles represents Blocked Bart's boundary?

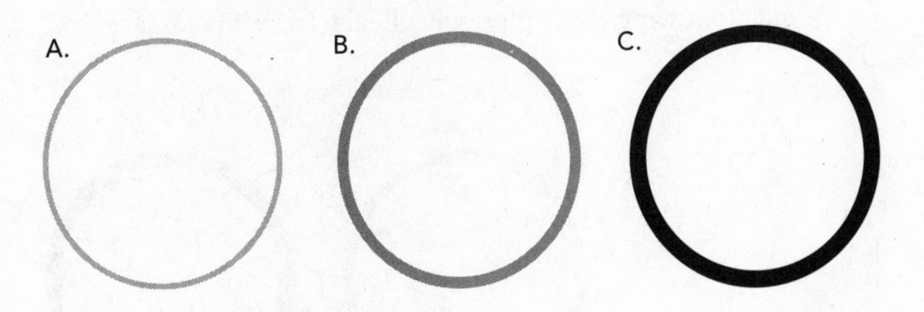

It's obvious, isn't it? Boundary C is thick and solid. Energy and information are trapped inside. Fresh information and the energy that can come from human interaction are bounced off the surface; they can't get in.

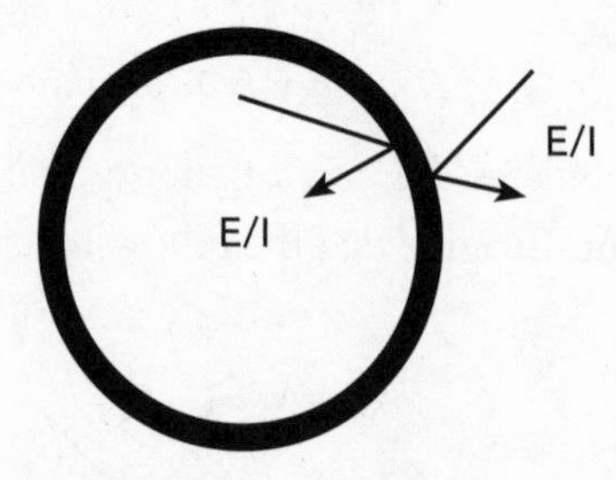

Imagine you were telling Bart about the joyful moment when you fell in love. How far would you get before your own energy petered out? Imagine having a discussion about a new candidate for president — a truly honest, brilliant, brave, capable person with a good chance at rescuing our sad country from its dilemmas. Do you think you'd have much influence on Bart?

Contrast Bart with Leaky Lucy, a neighbor you meet occasionally at the grocery. Lucy tells you everything. You know her entire family tree, that her husband has a hiccup at the end of his snore, that she once stole a piece of toast from a hospitalized man's breakfast tray (she was very hungry). She tried to tell you about the funny shape of her child's turd, but you walked away as fast as you could, forgetting that you meant to buy a pumpkin.

Which following circle represents Leaky Lucy's boundary?

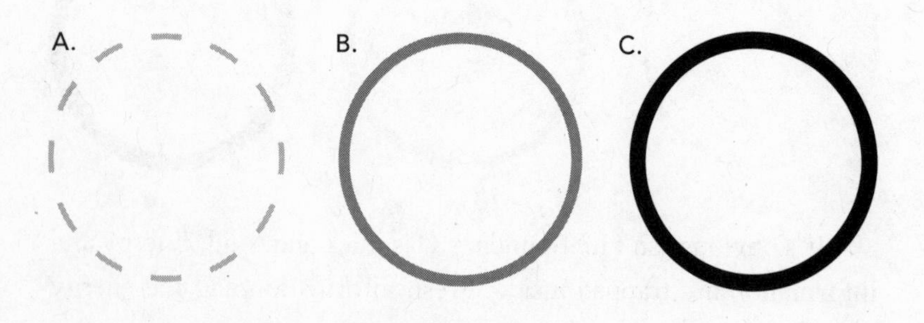

Clearly boundary A has holes in it.

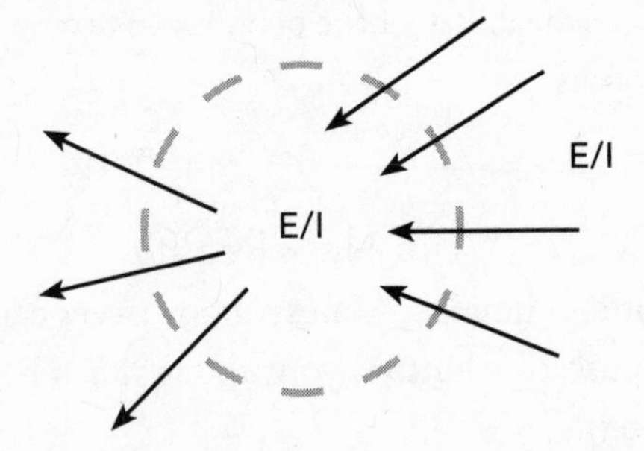

Lucy doesn't filter outgoing information, and her energy isn't contained. She doesn't discriminate between what's appropriate to the relationship and what isn't. And everything she's exposed to enters her conversation with others.

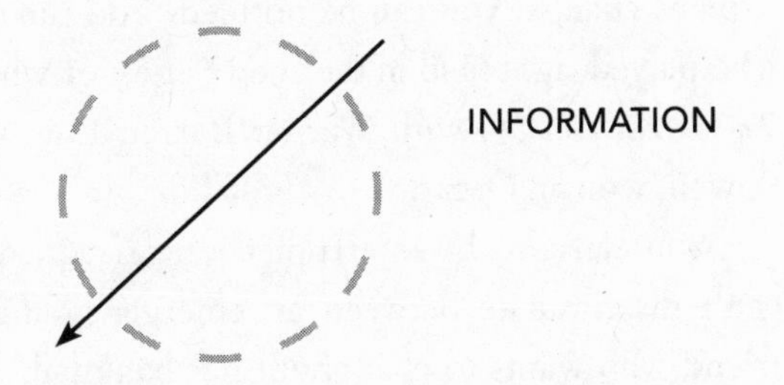

What would happen if you emailed Lucy a secret? If you told Lucy not to forward the email or tell anyone what it said, could you trust her to respect your request? (If she isn't able to keep her family's private information to herself, do you think she'll be able to protect yours?)

Consider your own family members. Which ones resemble Leaky Lucy? When you forward emails to your family, do your Leaky Lucys get copies?

Do you have any leaky friends? Are they among the recipients of your email about your lovely day with your friend who is in the witness protection program?

It's worth thinking about, isn't it? Give a day to thinking about the boundary permeability of the people you text, email, Friend, and visit in chat rooms.

The New Reality

The cyberworld is infinite. A message or tweet can live there forever. At one push of a button, your thought can rocket around the world in seconds.

Many of us start our days on our home computers, leave home carrying our smartphones and a tablet or e-book reader, get back on a computer at work or school, then go into our evenings with at least a smartphone. We are always connected.

There can be some comfort in this. You can always call for help. If plans change, you can be notified. You can settle a debate about who played right field in the 1960 Series, or whether the first film of *The Great Gatsby*, with Warner Baxter, Lois Wilson, and William Powell, won an Oscar.

You can also be interrupted on a frequent basis. Your phone can't discriminate between an emergency message and a boring friend who wants to chat about her hangnail. Your live experience with the flesh-and-blood friend sitting across from you can be punctuated with texts, calls, and instant messages (IMs), which will affect the boundary of your friendship.

You can feel wired in this wireless world. You are always available to everyone.

This constant connectivity has created a boundaryless world. Yet we are not its prisoners. Each of us can calm this chaos for ourselves.

Chapter 2

BOUNDARY FLEXIBILITY

Can you think of a circumstance in which it would be appropriate to toughen up your boundary so that it resembles Blocked Bart's boundaries? In which of the following situations would that make sense?

- Texting your daughter about your husband's great golf game
- Texting your daughter about your ex-husband's (her father's) adoration of his newly born son with his second wife

It's a foregone conclusion. The golf game gets a green light. However, spreading information about your ex's focus on a child in his new family could hurt your daughter, especially if he is a slapdash father to her.

What about texting that message to your sister, a mild Leaky Lucy? Can you trust that she won't someday mention this to your daughter?

Clearly, your boundary has to be different for each of these two subjects. You can sing like a birdie to anyone about the golf game. That message isn't personal or confidential. It won't harm you, your husband, or any recipients of the information.

Your ex-husband becoming an involved father after years of neglect toward his first-born child — that's a different story. It has the power to hurt your daughter, to create an issue between her and her half-brother, and to fan a flame in her relationship with her dad.

But I'll bet you've known people who have done that very thing — who have been indiscriminate about what they revealed that could hurt the recipient or harm an important relationship. Or perhaps they passed on potentially hurtful information deliberately, to vent their own anger or frustration.

So emotions, like information, can be boundaried or unboundaried. And a text or email or post on a wall is an easy, quick way to splash out an emotion that is spilling over. You might feel better or you might not, but once information is in cyberspace, it is beyond your control. You can't get it back. There is no Undo button once you've pressed Send.

When you tell someone something — on a phone or in person — it goes into that person's brain. It might be remembered, forgotten, buried under other incoming news, or passed on in summary form. Your long story about getting stuck in a snowbank on a deserted country road and digging towels out of the trunk to protect yourself from freezing, forced to eat the leftovers in your doggie bag and waiting till morning to walk to town for help, might be passed to a mutual friend as, "Ester is always forgetting her phone. She spent the night at the old Simpson place."

But when you text or email something, it's so easy for someone to pass it on verbatim, in all its glorious (or inglorious) detail. With a touch of a button, your embarrassing story about leaving your panties in the Walmart changing room can land in your minister's office, your nephew's dorm room, and his fiancée's purse.

Each time you email or post your information, you are trusting the boundaries, common sense, and discrimination of every person that information might reach. And it can reach nearly anyone on the planet.

You are trusting that Vengeful Val won't receive it from Leaky Lucy and forward it to your ex-husband, and that he won't send it to your ex-sister-in-law, who has never forgiven you for putting her pickle casserole at the back of the table at your wedding reception potluck thirty years ago, and that she won't send it to her cousin, who is interviewing you for a high-powered job — tomorrow.

Boundary Layers

Your boundaries are layered, and each layer can have its own degree of permeability. For example, you may have sensible limits around the information you share with others but Blocked Bart–type boundaries around your feelings, shutting them completely off from yourself and others.

You may have healthy boundaries around your food choices, but leaky boundaries around the clothes you pick, wearing clothing that is too revealing in your workplace. You may have thick boundaries with a malicious coworker, being quite smart about giving her nothing to use against you, but see-through boundaries with your daughter, taking on too many of the troubles she brings onto herself with the choices she makes. You may push yourself too hard and toil for too many hours, not putting boundaries around your work efforts, but have too thick a boundary around receiving help from others.

Your boundaries establish the highway of your life. They fence out certain options and let others in. They create U-turns, detours, accident sites, limited access passages, and speed lanes.

And you are the architect. You create the characteristics of your boundaries via the choices you make.

We each got where we are by making choices. The good news is that you can get yourself to a wonderful new place, and regain authority over your life, by building healthy boundaries.

Chapter 3

BUILDING BOUNDARIES

Boundary building can work from outside in or inside out. You can build boundaries with conscious effort and this will change you on the inside — or you can change your insides, and boundaries will automatically form. You can set deliberate limits to protect what you care about — or they will spring up by themselves when you focus your whole self on what matters to you most.

We'll work on this from both ends. You'll practice setting deliberate boundaries, and you'll also practice managing your focus so that you automatically improve your boundaries.

Boundaries are influential. As you become more appropriately boundaried, people will automatically treat you differently. The more effective you become, the more effective people in your sphere will become in relation to you.

Boundaries are isomorphic. When you hold a boundary in relation to your family, your family as a whole will come to hold a similar boundary in relation to you.

(The term *isomorphy*, as used here, comes from the cutting edge of psychological theory. In a hierarchical system, there are similarities in how the boundaries function. A family is a hierarchy. A child — an individual — is a member of the sibling subgroup, which is contained in the nuclear family, which is contained in the extended family. In an undemonstrative family, there may be an unstated, but understood, boundary against hugging. The sibling subgroup, exposed to a school culture in which hugging is commonplace, will bring the more permeable boundary back to the family. As they begin hugging parents and grandparents, the altered boundary will enter the family's culture, and the family boundary will shift.)

Boundaries can be visible, like a fence. They can be verbal, as with a very firm "no." They can be demonstrated by what you do or don't do. They can be created by focusing energy. And they can be communicated by what you stop another person from doing.

Hester stands next to me at a community chorus rehearsal. From my first moments of being in the chorus, Hester has treated me as if I were new to music, leaning over to inform me that the pound sign indicates a sharp or that *piano* means soft. I play three instruments and have studied music since I was a tyke, but she assumes that I need basic instruction.

This is typical of countless small interactions that occur daily with people on our periphery who, like all of us, wear their self-image on their sleeve. Without knowing it, we communicate how we want others to view us. Meanwhile, we might be trampling on how the other person wants to be viewed.

In many of her interactions, Hester broadcasts her belief that she knows better. At first, I tolerated her interruptions, but soon I noticed that her behavior was starting to influence me. Her interruptions were distracting me and arousing my competitive nature, so I focused less on the music and made some mistakes. I wonder if others start to slip in their own performance around her.

Since I only saw Hester once a week for a couple of hours, it was tempting to let the whole thing go. But I decided to act before I got irritated — a likely eventuality that could give me more to deal with and even taint chorus for me.

I realized that I'd have to set a boundary. This would protect a pleasurable pastime and give me more energy (and less distraction) for singing.

I had some choices:

- I could mirror Hester and feed her behavior back to her, explaining to her that the dot with the eyebrow is a fermata.
- I could compete with her by being more alert, more accurate, and more responsive to the director. This choice had the potential of escalating the situation, of putting pressure on both of us that could cause mistakes.
- I could reveal more of myself, letting Hester know I'm a musician, which might indirectly set a boundary by correcting her misconception about me.
- I could seat myself very firmly in my own energy so that she'd be diverted to some other poor soul in the choir.
- I could say something to her directly.

I decided to try a three-tiered approach. I'd start with a droplet of information about my musical experience and then be very solid in my own energy. If these didn't work, I'd say something.

The first two didn't work, so I said softly that I appreciated her experience, but that I found myself being distracted by her comments. If I wasn't sure of something, I'd ask her, but otherwise I'd rather stay focused on the music and the director.

She drew back. She looked startled.

Fear of such a reaction is one of the reasons many nice people hesitate to set boundaries. They don't want to hurt the other person, so they endure being hurt themselves or they let something precious to them be tarnished.

I sensed that Hester was offended. Many of us would be, but she had an opportunity here, too — to learn something about the effect she has on others, to see herself more clearly and think about the needs that prompt her behavior. This is the opportunity afforded by any appropriately drawn boundary: It gives important feedback that can spur the other person's growth.

Regardless of what Hester did next, my challenge then was to stay comfortable within my new boundary, to not weaken it to make her feel better. When I saw her shocked reaction, I might have been tempted to soften it or say, "Oh forget it. It's not that bad." I could have rushed in with a lot of words to build her up or put me down or throw fog into the communication. And I would have undone my positive effort.

So I did none of these. I simply stood quietly, letting my words sink in.

When you build a boundary, don't tear it back down because of a person's reaction — or your fear of how she might react. Let the boundary stand. Let the other person be responsible for her own reaction.

> RULE OF THUMB
>
> Every time you set a boundary, you'll get better at it, and your life will improve.

That's what I did with Hester.

My responsibility was to set the boundary clearly and thoughtfully. What she did with it was her responsibility. She'd learn or not learn, change or not change, be victimized or angered or surprised or grateful. That's her own work.

I'd have created more accountability for myself if I'd let the problem go on too long and then set the boundary angrily. If I'd said, "Hester, damn it, stop correcting me" — especially if I'd spoken in an angry voice that others could hear — then I'd have to take care of the consequences of venting instead of communicating.

It's also my job to maintain my boundary in the future. If Hester respects my boundary for a time and then forgets or slips up, it will be my responsibility to say, "Hester, please don't. Please remember what I asked."

If she then continues to push it, regardless of my mild comments, I'll have to set a stronger one. I'll have to say firmly but discreetly, "Hester, stop correcting me."

It's my job to continue to protect my pleasure and my energy. If my comment is clear and honest, and not emotionally loaded, I'll be participating in creating a clean relationship, regardless of what Hester does with it.

The steps I used are these:

> What you put up with is what you end up with.

1. I noticed a boundary was needed. (Hester's comments were intrusive.)
2. I decided what I wanted. (For her to stop correcting me.)
3. I scanned my options for communicating what I wanted.
4. I set my first boundary thoughtfully, clearly, and respectfully.
5. I let her reaction be her business.
6. I supported my boundary. I didn't weaken it to try to fix her reaction.
7. When she continued with the same behavior, I reaffirmed and strengthened my boundary by being clearer and more direct.
8. I brought my attention back to myself and noticed how it felt to protect myself.

EXERCISE: Setting a Verbal Boundary

Now it's your turn.

Let's start with a single project. Think of a situation in which you need to set a boundary. It could be with a person to whom you always say yes when you want to say no. It could be with someone who peppers you with questions that you don't want to answer. Perhaps someone takes things from your office without asking.

Now answer the following questions:

1. What is the situation?
2. Until now, have you:
   - Endured it?
   - Lashed out?
   - Retaliated?
   - Felt helpless?
3. How has that worked for you?
   - Has anything changed for the better?
   - What has it cost you to put up with it?
   - The last time it happened, how did you feel afterward?
   - Do you agree that a boundary is needed? If so, continue.
4. What do you want? Either write it below or say it clearly to yourself.

   I want ______________________________

   ____

   ____

   ______________________________________.
5. What are some statements you could make to the person that would set a boundary?

   (Examples: "I'm going to say no to that, Ellen. I'm not

going to do that." "Evia, please stop asking me questions. Tell me something about you." "Max, stop taking things from my office. Email me if you want something, and please wait for my reply.")

Write some of your own possible statements below:

____

____

____

____

____

6. Which of your statements is clearest, most straightforward, and most direct? You can start with that one — or, if that seems too harsh, pick a milder one that still states specifically what you want.

When you're clear about what you want to do, follow these steps:

a. At the first opportunity, make your statement to the person involved.
b. Don't let the other person's reaction be your business. Notice how it feels to leave it with that person.
c. Notice your impulses. Are you tempted to take back or weaken your boundary? Are you tempted to explain it three different ways?
d. If you do weaken it or take it back, notice how that feels. If you hold the line, notice how that feels.
e. Notice whether the other person's response triggered a reaction in you. If so, what was your reaction? Did you act on it in any way?
f. Notice how it feels to protect yourself, to safeguard your energy or something else important to you.

g. Once you've set a boundary, monitor it. If the other person starts slipping and reverting to old behavior, set the boundary again. Use a stronger statement.

Resources

For more basic information about boundaries, read my book *Boundaries: Where You End and I Begin*. For lots of examples about how to set boundaries in specific situations with specific people, read my book *Where to Draw the Line*.

PART II

BOUNDARY VIOLATIONS

A boundary violation is an action (or failure to act) that weakens or breaks a boundary, harming the entity inside it. There are two types of violation:

- Intrusion violations
- Gap or distance violations

An intrusion violation penetrates the entity, creating a breach or interruption.

Examples:

- A bullet breaking through the boundary of the skin can kill.
- A man who hits his wife shatters at least two boundaries: He bruises her body and he ruptures the boundary of the relationship. If any children are watching (or listening), their boundaries of familial safety are also breached.

A gap violation is created when one fails to act or respond when an action is called for.

Example: A parent who doesn't reach out to his or her miserable child is violating a care boundary, and both the child and the parent-child relationship are harmed. There's a gap between the normal emotional need of a child and the parental care that would promote the child's healthy emotional development.

The missing deed puts distance into a transaction and will, if repeated, move a relationship away from intimacy.

Chapter 4

INTRUSION VIOLATIONS

Crossing a Boundary

When a certain friend forwards a bunch of emails to me all at once, I know two things: that she's on vacation, and that I don't have time to read all of them.

When you get inundated with a friend's forwards, you can just delete them sight unseen. But if you are a member of a group that will digest them, and some of the forwarded content enters your group's culture, you'll be left out. While your friends laugh themselves silly after offering you an acorn, you'll be in the dark. You won't know about the video that showed the squirrel putting acorns into the park drunkard's comatose mouth.

Intrusion violations can be physical, emotional, or veiled. They all cross a boundary into your inner space. If your boss shoves you, that's a physical violation. If he screams at you, venting his frustration inappropriately, that's an emotional violation. Copying you with every work-related email he sends to anyone, "just to be safe," is a veiled violation. We feel the impact of a veiled violation, even if there's no harmful intent.

If you ascertain that something exists by its effect, rather than by observing the thing itself, that thing's existence is veiled. Scientists sometimes discover the existence of a planet or a star, not because they can see it — it's invisible to them — but because it exerts gravitational pull on heavenly bodies they *can* see.

I have a friend who sent daily hailstorms of emails after she retired. They filled up my inbox, and deleting them took time I didn't want to spend. They were all high-quality emails, too. They included inspiring pictures of the planet or adorable pictures of animals and babies. There'd be a gimmick showing how to open a locked car, or a cute joke about retired women. Despite the fact that my friend had good intentions, my time and energy boundaries were crossed, creating a veiled violation.

One email with a picture of a cat holding a baby has little impact. Thirty emails exert a gravitational pull. We can feel the effect even though the individual forwards are harmless.

I set a boundary with my friend. I said, "You have wonderful forwards, but I don't have time for them. Each week, pick your favorite and send that one to me."

She respected my boundary. For a long time, she sent me one forward a week. Now that boundary is slipping; she's sending me three or four email forwards a week. I'll have to remind her and reinforce my boundary.

Setting Email Boundaries

I dread opening my email program every day. Even though my spam filter does a good job of weeding out junk, I'm still forced to momentarily view the agendas of many other people in the middle of my own agenda-packed day. When it's over, I'm relieved, but my energy level has decreased. I don't get nearly as much done as I might have before I got diverted by sifting the wheat from the chaff. And there's a lot of chaff.

It's tempting to not check email at all, but I don't want to miss an invitation from my cousin, the latest news from my best friend, or an important request from my literary agent.

This is a veiled intrusion violation caused not by one person but by the combined actions of many people, most of them strangers, sending out emails willy-nilly to further their own purposes. And this problem is here to stay.

> RULE OF THUMB
>
> Here's how to set a Rule in Mac Mail:
>
> - Open Mail Preferences.
> - Choose *Rules.*
> - Choose *Add Rule.*
> - In Description, name your rule (example, *Pills*)
> - Select *Message Content* in the first drop-down menu on the third line.
> - In the third box on that line, write *Viagra.*
> - Choose *Delete Message* as the action desired.
> - Select *OK.*

We must set boundaries to protect ourselves because the bombardment will continue. Here are some options:

- Use an email program that has a feature for identifying junk mail and gives you options for deleting these items or filing them separately.
- Choose a server that allows you to block particularly pesky email senders.
- Check to see if your email program lets you set rules (see sidebar).
- Notice whether most of your forwards are from particular friends. Decide how many emails (and what kind) you want from each forward-happy friend. Call those friends and set that boundary, then follow it up with an email. Examples:

    - Please choose your favorite forward and only send one per week.
    - Please only send me cartoons about gardening. I'd prefer not to receive forwards of any other type.
    - Please don't send me any forwards at all. I simply don't have time for them. But I want to always have time for *you*.
- Decide if you want only a particular category of forward, or no forwards at all, from anybody. Make that request in an email that you send to everyone on your email list.
- Decide whether having emails pinging in one at a time interrupts your concentration. If so, close your email program and only open it when you are ready to turn your attention to emails.
- Notice what happens to your energy and focus after you check your email. If your energy decreases or your focus gets scattered (indicating a boundary violation), wait to check your email until the end of your productive day.
- Set a time slot in which you check email and let everyone know when it is. Example:

  Dear friends,

  Not wanting to be interrupted while I'm in my creative mind, I've decided to check my email only three days a week, after 4:00 PM. If you need a quick response from me or want me to know something outside those times, please call.

- If you need to be able to receive certain emails from certain people to do your work, set up a separate email address for just those people. Use either a separate email program that checks only that one account or, if you have a smartphone, leave only that account open.
- Remember to send those people your new separate address.

Let them know this is a special, private address, just for this work situation. Ask them to please not forward any of your emails to anyone outside this work group or to give this address to anyone else.

You may have noticed that I made a distinction between a blanket email that cancels all forwards from all friends, versus a phone call setting boundaries with a couple of forward-happy friends. This kind of discrimination is required of all of us, now that the modes of communication have multiplied.

Setting a boundary that isn't personal — such as a boundary that blocks all forwards in one fell swoop — is a communication from you to everyone. Individuals aren't being singled out, and you can communicate this in your email.

Hi friends,

My email inbox is so full each day, I've decided to eliminate forwards. So please take me off your email forwarding list, and don't forward anything at all to me, no matter how cute, funny, politically correct, or awe-inspiring. I'd rather hear from you personally anyway.

Please respect this boundary. It's important to me. You will be supporting my work and my peace of mind.

Thank you,
Harried in Houston

Setting a boundary with a single friend is a different matter. How would you feel if the following email came to you?

Twila,

Please stop sending forwards to me. I'm just so busy I don't want them cluttering up my mailbox.

Groucho

Groucho sounds grouchy and a bit rude. You might be hurt or offended if you received his email. Would you be more or less likely to comply with this request than the one above? Could this missive have an effect on your relationship?

On the other hand, a phone call can be a good approach for a one-to-one request. It lets you hear Groucho's tone of voice. You can ask questions. He can hear your response and clear up any misunderstandings immediately. It will be a dialogue:

"Twila, Groucho here."

"Hi, guy."

"Twila, I'm up to my eyebrows in emails. Could you do me a favor? Could you limit your forwards to one a week? Just pick the one you like best."

"Sure, Groucho. Don't you like them?"

"It's not that. You send such great forwards I can't resist checking them, and before I know it, I've lost an hour. I'm just too busy and I don't have any self-control. So if they don't even come to me, I won't be tempted and I'll have my hour back."

"Okay, Groucho, I'll be glad to protect you. Are things okay between us?"

"You bet your life."

As you can see, Groucho set his boundary, protected his time, and reduced clutter — and you remained buds.

How Communication Tools Became Instruments for Boundary Violations

In the olden days, we had two forms of communication: letters and face-to-face meetings. Somebody realized that mass mailings could generate business, and junk mail was born.

The telephone arrived. (The telegraph came first, but there wasn't one in every home, so it didn't get a chance to bother us.) This speeded up communication considerably. Unlike letters, telephones allowed for direct dialogue, and each party on the line could hear inflection, tone of voice, and, often, emotion.

For many years, a phone was operated by dialing a round disc with holes in it. When we called someone, we dialed the phone. We still use that term, even though today's phones have buttons or number screens, and we actually push or tap a friend's number rather than dial it.

Eventually, most homes had one telephone with a single number. Only people who knew you called you, and when the phone rang, you either knew the person on the other end, or you knew the person they meant to call. You may have had something to tell them, even if their call was intended for someone else. (In my small town, this is still the case.)

Then a businessperson realized that people could be hired on the cheap, planted in a chair, and given the job of calling numbers all day, delivering a spiel if anyone answered. And telemarketing became an industry.

The first answering machines were actually invented soon after the telephone, but they didn't come into demand until after — you guessed it — telemarketing. Now voice mail is integrated with your phone system, but in the 1980s, message recorders were a separate box that sat next to your phone (which was plugged into the wall). Answering machines were not only convenient for receiving messages when you weren't home but also a way to circumvent those irritating telemarketers — a response to the boundary violation of strangers interrupting your peace by calling you with their agenda.

Then email flew in. This made possible the quick dissemination of identical messages to many people at once. Replies could be immediate (much faster than waiting for the postal service). We discovered that addressing an envelope and finding a stamp was a lot of bother. We had a written record of the new bowling time, and we could refer to it if later we forgot. Our fancy stationery yellowed on the shelf.

Email was so easy, we began connecting with long-ago released people — old classmates, ex-husbands, army buddies long out to pasture. We could reach out to our Belgian cousins as easily as to a neighbor. We felt jubilation at the sense of connection. The world felt smaller. It seemed like a preamble to world peace.

In short order, commercial minds saw the potential of email. This was a heretofore unknown way to market inexpensively to the masses. At little cost, hundreds of thousands of people could be made aware of your amazing widget in the blink of an eye. The downside (for us) was that there was no way to distinguish a commercial email message from a personal one. We hadn't liked it when telephones began to be used by solicitors, so we were pleased to screen callers by using answering machines and caller ID. But now we were faced with the same problem all over again.

As we got savvier about email, marketers got savvier about the subject line: "In response to your inquiry," "Urgent Message from Your Bank," or "Your Credit Rating is Sinking." The slick linearity of early email got tangled into a Gordian knot.

Texting, the next development, became an interesting amalgam of email and phone. Text messages arrived by phone. They were written. They could reach you even when you were away from your computer. They could be sent to multiple recipients.

Commercial messages have now begun to arrive by text (and you have to pay for them), but there is a hitch: Solicitors have to get your number from somewhere. There's no convenient cell phone directory (as of this writing). You have to be tricked into supplying your number — by a friendly, innocuous mall person with a clipboard who will give you two free tickets to the roller derby if you fill out their little card, or by a kiosk at the fair that offers coupons for free popcorn if you answer its electronic questions.

Filters

You can protect yourself to a certain extent by getting on various Do Not Call or Do Not Contact lists (go to www.donotcall.gov). Certain organizations — such as charities, political organizations, and telephone surveyors — are exempt, so if you receive a call from any of these, you can tell them to take you off their list. Many states also have Do Not Call registries.

A company with whom you have a business relationship can call you even if you are on the registry. This becomes relevant if you fill out that sweet mall lady's form, or a teeny coupon for a free DVD drawing at the online store for Kaleidoscope Media Associates. KMA also owns thirty other businesses that engage in telemarketing, and since they now have a business relationship with you, all thirty can call or email you.

With great irony, a recent scam involves someone calling you, saying they represent the Do Not Call registry, and offering to sign you up. Does it ever end?

To protect yourself from junk mail, junk email, and credit card offers, go to www.ftc.gov and choose *Consumer Protection.* The Federal Trade Commission lists resources for shutting off the flow of unwanted advertising, and the links to Do Not Send Me Stuff.

From Typing to Tapping — Necessary Distinctions

Once we had one phone number per household. Now we have many. These days, when we call a friend, we may call three numbers and leave three messages. The number of phones has increased — efficiency, not so much.

We have to factor in what we know about a person before we call. Kyra never uses her cell phone. Jax is always connected. Tallie gets her messages. Jake may ignore messages for a week. The Simpsons don't use their cell phone when they are home, and don't call home for messages when they are away.

The plethora of modes of communication requires us to make distinctions about which mode to use when. In part, modes are generational. My grandmother wouldn't have considered phoning a thank-you; she always sent a note. My mother doesn't use a computer. When I gave her a cell phone, she couldn't grasp that she had to put the thing to her ear. She held it in her lap. I'd be yelling fortissimo into the phone, "Mom, put the phone to your ear!" and from a distance I'd hear a weak, "Hello? Hello?"

My friends use email more than they phone. They can send an email at a time convenient to them, and I can read it at a time convenient to me. My best buddies and I send group emails. Everyone keeps up with what everyone else is doing.

My younger friends text. They pepper each other with texts. I've had to learn to text in order to meet them where they operate most comfortably.

The Newer Kid on the Block: Texting

Text messages create additional issues. They require dexterity. If you have small thumbs, or a smartphone with a sideways keyboard, you can train yourself to tap a fairly rapid hunt-and-peck message.

As with email, you can send pictures. You can reply immediately. You have a record of the string of messages.

A text message lets you be in control of the dialogue. In fact, you can make it a monologue. You can send a text when you don't feel like talking, when you don't have time to discuss the subject, when you don't care about what the other person wants to say to you in reply, or when you want to command an action without back talk. You can receive a text when you are at a business meeting, while giving a speech, while talking on the phone, in a loud stadium, or at a movie. A text is a convenient way to cancel an early-morning walk with a friend, without waking up the household.

Recently, a younger friend and I were watching the same TV

program in our separate homes. Throughout it, we texted about our reactions to the program, as well as about an issue she was dealing with. It was great fun, and a way to share an experience from separate places. It didn't bother anyone else who was in the room watching the same show.

Texting is very useful for messages such as these:

Mom, I'm home from school.
Hon, tied up in meeting. Will call later.
Harry, dog is in yard. Fetch.

Texting is a great way to send and receive messages without anyone around you knowing, in a context in which using your phone would be disruptive or get you in trouble.

Your child may not answer the phone or check voice mail, but a text is irresistible. The message is right there, graspable in a few seconds.

If you just want to say something and don't actually want to talk about it, you can send a text and then turn off your phone or ignore any responses.

You can keep in touch with your mate throughout the day, sharing the highlights in absentia. By sharing your separate realities, you feel closer — as long as you still talk when you rejoin each other.

Disadvantages of Texting

Texting actually takes quite a bit more time than talking. In terms of the amount of information imparted both ways, a brief phone call is often more efficient.

If you ask multiple questions in one text, you're likely to get an answer to some, but not all. You'll have to re-ask the unanswered questions. You can also text one question at a time, but this takes more time and requires a good memory (so you don't forget your other questions).

Autocorrect can make for some interesting messages.

Hi just PDA testify. Hoping you see well. Back to Marie die second time toesy. Keep forgetting Rocco.

(Hi, just practicing texting. Hoping you are well. Back to market for the second time today. Keep forgetting broccoli.)

Furthermore, it is even more dangerous to text while driving than it is to use a phone. Even picking up your phone to read a text can put your life and the lives of others at risk. Every year at the state fair, the state patrol has a mangled vehicle on display. These used to represent the last journeys of drunk drivers, but in recent years, the cars have been totaled by little bitty phones. (Did I mention that an incoming text is irresistible?) The phones didn't survive. Neither did the drivers.

Mixing Media, or Tepid Texter Meets Mighty Messager

A miscommunication may be caused simply by mixing media: responding to a text by calling the other's home land line and leaving a voice mail, or answering a complicated emailed question with too brief a text. For example, a tepid texter, Trudy, who is friends with an intrepid tapper, Tony, doesn't answer his text messages because she prefers email. Unless he understands this, and they both negotiate how they are going to handle various types of messages, too many one-sided communications may push him away.

Leaving one complicated voice mail isn't necessarily a boundary violation, but several can show a lack of awareness of the recipient's situation. On the other hand, not answering messages because you don't like the other's medium may cause more work for the sender. Repeated communications that cause extra work for either person can weaken a relationship boundary.

If you care about a relationship, explain which devices work for you and plan how you will communicate if you each have different preferences.

Boundary Equity

Boundary equity promotes a healthy relationship. Communications that show a lack of attunement or are one-sided will weaken the boundary of intimacy. I'll give you an example; although it involves verbal communication, it could just as easily have occurred through a phone call or email interchange.

The other day I was at a restaurant, and I could hear a man at a table behind me. He talked nonstop. His companion said nothing. This went on while I ordered, waited for salad, ate my salad, waited for the main course, and ate the rest of the meal. I couldn't resist; I had to peek.

He was involved in his monologue. He gestured, he laughed, he looked around. The woman with him looked down. She slowly ate her meal, her eyes on her food. She gave him no encouragement. She did not reply, ask questions, or enter the stream of words with thoughts of her own.

Would you say the boundaries were equally permeable in that relationship? Were both people thoughtful to about the same degree? Were they both having a great time?

One way to look at boundary equity is to weigh the amount of extra effort you must use to straighten out an ambiguity created by the other person. Is the sender communicating in a way that makes more work for you?

Cryptic Carol doesn't complete sentences. She starts three sentences and finishes none of them. I can intuit her intent and say, "Did you mean to say the turtle isn't in the pond?" I can retaliate and say, "No," knowing that my answer is as ambiguous as her non-statement. I can wait, leaving her communication as her responsibility, and then I have the work of managing my irritation.

I've noticed that, in a group, our friends rush in to fill the gap. They all work on possible sentence endings and she sorts through them. "No." "Uh-uh." "Yes, that's what I'm wondering." It's an interesting dynamic, and it makes more work for her friends who end up communicating for her.

(Of course, if a person's ability to communicate is compromised by a physical condition, such as a stroke, that's a different situation entirely. Extra effort on your part to understand is a gift given out of love or friendship.)

Cryptic Carol brings her brevity to texts and emails, being either vague in her pronouns or incomplete in her messages. To get the name of the restaurant or to discriminate whether it's her older or younger daughter who just got engaged, you have to text back.

If Blocked Bart doesn't answer reasonable questions sent by any medium — text, email, or phone — you may end up trying harder. You might ask again, or you may not know what he wants and end up fixing two main dishes, not knowing if he's still a vegetarian. His boundary is too closed, and yours is open. This difference creates a dilemma that can weaken a relationship.

Of course, there are times when a message shouldn't be answered. "Are we walking tomorrow?" should be answered. "Are you wearing your new bra?" is appropriate only in limited circumstances. The first question, not answered, causes work for the asker. The second question, if it isn't appropriate, makes work for the responder.

Interchanges that cause one person to work harder will have an immediate effect on the boundaries of the person doing the extra work and, over time, will affect the boundaries of the relationship.

Not answering — or not explaining that one doesn't check email or doesn't like to text — can create a gap violation. Using an inappropriate device for the situation — or causing more work for the other person — is an intrusion violation. Intrusions will always, ultimately, be walled off.

Match the Message to the Medium

Don't switch communication modes without letting the other person know. People expect to receive a reply in the mode in which they send their message. If they text you, they'll expect a text in return, unless you tell them otherwise. If they call, they'll expect a return phone call. You can confuse things, and possibly create a gap in communication, if you answer a phone call with an email, or a text with a post on Facebook.

In a personal emergency, call. It's quicker. You may not be able to think clearly enough to realize what information the other person needs. They will be able to contribute common sense, help, and direction to the situation. They can call the other people you care about.

In an emergency, a text message or email leaves the other person hanging, especially if you, caught up in the demands of the crisis, aren't able to respond when they reply. This can delay personal help reaching you. The other person may not have the logistical information they need to reach out to you or your loved ones. And, of course, they may not be checking their email or texts.

If disaster strikes nearby but you are not involved, text. When the I-5 bridge collapsed just after rush hour in northwest Washington, cell towers became overloaded with the thousands of calls that flew from onlookers and people sitting for hours on the interstate. Workers on the scene, agencies, officials, and journalists kept losing their signals due to this high phone usage.

Therefore, texting is the method of choice in this important situation. In a disaster that doesn't affect you directly — a bridge collapses, a bomb explodes at a large gathering — text your people to let them know that you are okay. This leaves cell towers available for the many responders who are directly involved in handling the disaster.

You may want to talk to people about your experience, to tell them how close you were, how long you were stuck on the highway, what you saw, how you felt; but resist the temptation to call, and instead communicate these things by text.

We're all still learning what is appropriate in each situation. This is part of our training, noticing that with a rich array of choices for communication, we can pick one that does the job we need it to do, while respecting and supporting the needs of the professionals in the same situation.

Of course, if you are directly affected by a disaster, do use your phone if you can, because your loved ones need specific logistical information from you, and they will have questions for you. On the other hand, if you are near the disaster — in the town or the area that shares phone lines or towers with the affected scene — but not in it, text, email, or use social media, but do not phone.

For a change of plans at least a week in advance of the event, email or text, asking recipients to acknowledge that they've received the message. Call any who have not confirmed within 3 days of your notice of change.

For a change of plans less than a week before the event, call, unless the list is so large that you need to start with an email or a text. Ask for confirmation of receipt of the message by the end of that day, and call all who haven't replied.

For a quick check-in, text. It's a slight interruption for the other person and settling for both of you. To quickly say that dinner is ready or that you'll be late, texting is useful.

For a complicated, non-urgent message, email. You'll be able to spell out all the pieces, and the other person can sort through the details at their own pace.

To work out an issue, call or meet. I'm surprised at the number of couples who try to resolve a conflict with email or text. This makes a muck of misunderstanding. Texting requires too much time to spell out a complicated idea, and reducing a message to a few lines creates ambiguity. Your quick and innocent quip may be read as snide or mean. His silence may seem resistant or obstructive. Misinterpretations grow rampant, complicating an already volatile situation.

With text messages, your ability to discriminate between reality and unreality is hampered. You may respond to your own speculation about what he is thinking, instead of his actual thoughts. You could end up fighting about a nonissue, weakening your intimacy boundary.

My friend Maurine recently had major surgery. Her partner had designated another person to call their friends so we'd all quickly know the surgery went well. Then, later, her partner emailed us details about the results, the recovery, and what support was needed. It was a perfect combination of modes of communication.

Apprentice blacksmiths, cabinetmakers, and farmers are taught, early in their training, to use the right tool for each job. The same is true for communicating with others — especially when an electronic tool is involved.

Are These Boundary Issues, or Just Good Sense?

A communication problem becomes a boundary issue if you lose your energy, if you end up with inaccurate information, or if it weakens the health of a relationship.

Remember that a boundary does several things at once: it defines an entity; it regulates the flow of energy and information; it protects wholeness; it promotes autonomy. An intrusion threatens each of these aspects.

Emotional trauma is defined as a distressing experience that a person is unable to escape or stop. Someone else is causing the experience and has control over the situation. The intrusion of an urgent text or email that creates concern in its reader, while rendering her helpless, meets the definition of trauma. It is, thus, a boundary violation for the recipient. When this happens repeatedly, the boundary of the relationship will be affected as well.

Think of an alcoholic who always has some crisis. At first, you respond. You care; you try to help. When your efforts bring about no positive results, you have spent your energy, received little in return, and been diverted from your own interests and life focus. If you are also plummeted into despair about your loved one or your fractured relationship, you are being traumatized. Eventually, you have to turn a deaf ear to restore your personal boundary and regain your own peace of mind. The boundary of that relationship will therefore change. The potential for intimacy has, of necessity, been walled off.

Device boundary violations may be more subtle and slower acting, but they still cause damage and, therefore, must be dealt with.

Chapter 5

ENTHUSIASTIC INTRUSION

Forwards

Email forwards used to be an easy way to pass on to friends something that tickled your funny bone or made you say, "Aw." Now, forwards are an effortless, instant way to commit a boundary violation, with certain exceptions. They are like junk mail from you to your friends; they clutter.

> RULE OF THUMB
>
> **Forwards**
>
> - Check first.
> - Get permission before punching *Send.*

In your intimate circle of friends, your forwards may be part of the group culture, but for anyone else they are at best a pseudocommunication, at worst an intrusion. (At the end of this chapter is a tool you can use to define exactly what limits your friends and acquaintances want you to respect; see pages 50–51.)

Forwards include several categories. Let's look at each and how it can be a boundary violation.

Jokes and Cartoons

These are the most innocent of boundary violations. Something made you laugh; you want to share it with others. If they've given you permission, your immediate circle of friends may also enjoy the joke, especially if it is apropos of your group culture.

With fairly close friends, the joke may be relevant to a point of view you all share. In this case, after you have their permission, a forward may fit.

With anyone else, jokes interrupt — particularly if you are forwarding them to your business connections. (More about online business behavior in chapter 19.) All the fun of the joke will be leached away if someone sees your forward and groans because it's just one more item that has to be deleted among thirty other unwanted and unasked-for emails.

Videos

The problem with a video is that it takes much more time to view than the quick glance required by a cartoon. A video link takes your recipient to another site, where it has to load — and most are at least several minutes long, some much longer.

Of course, your recipient has the option of deleting it unseen, but, depending on their relationship to you, they may feel obligated to at least check it out.

If you very much want to forward a video (or a link to that video), send your friend an email about it with the link at the bottom. In your email, describe the general subject, why you liked it, why you are sending it, and how long it takes to view it. It's only fair for you to do the up-front work, rather than make your friend do it.

Chain Letters

Most chain letters are a subtle way of promoting a cause. Young people won't remember the olden days when we had to write out all chain letters by hand (and send a dollar to someone in hopes of making a fortune). When copy machines arrived, we were thrilled at how easy it was to make twelve copies in a couple of minutes. Still, addressing the envelopes took enough time to keep us from sending chain letters willy-nilly.

Email let us embrace the ease of copying *and* sending, and chain letters flew wildly through cyberspace. Most of them now are not for money but for some worthy endeavor — a cancer cure, prayer, positive visualization — or at least that's what they appear to be.

At the least, chain letters are time eaters, especially when added to the pile of other forwards that have to be dealt with. And they encourage you to pass on that same time eater to all your friends.

In addition, a chain letter may contain a web beacon that isn't friendly, telling someone you don't know that you opened the email and who you sent it on to, marking you as potentially gullible and also giving the unknown original sender your email address and the addresses of everyone to whom you forwarded it. Now they can email you directly with their solicitations for an amazing fruit that will make you lose weight or for the world's best tweezers. After all, you don't actually know who started that chain letter (or that cute picture).

So if you groan when you get a chain letter, don't send it on.

RULE OF THUMB

Copy and Paste instead of Forwarding

To resend a message that fits the criteria in this chapter, copy the content and paste it into a new email.

Instead of automatically copying a link, just type the domain name. For example, type www.example.com, letter by letter, into the fresh email. This way, you reduce the possibility of passing on bugs or beacons.

You'll only be creating a chain groan as your friends get stuck with the same issue — whether to delete it and possibly ruin a chance at world peace or to burden yet other friends.

Information about Social and Political Causes

Now we're moving into categories that are your agenda, but not necessarily your friend's. Either you are trying to persuade someone to your point of view, or you want someone to take action on a cause you believe they share with you. You are, therefore, asking for more time — either to read a lengthy attachment or to go to a website and get involved, and possibly to mull over your points and positions.

As with other Internet violations, the problem is that it is so easy to do this. It takes you only seconds to send something that will cost the other person minutes.

Think about this: Would you go to that person's house for the express purpose of involving them in your concern? If you went to dinner with them, would you bring this material and present it to them there? How many minutes would you take to talk about this cause as you share a meal?

If you would make these efforts, then go ahead and make them in person. Go to your friend's house or take your friend to dinner, and present your case. If your friend doesn't live nearby, send a snail-mail letter about it, or call to discuss the issue.

But if you wouldn't do these things, look at why you wouldn't. Would it seem intrusive? Could it negatively affect the relationship? Would it be inappropriate, given the nature of the relationship? If the answer to any of these questions is "yes," then it's just as intrusive or inappropriate online.

Post your views on your Facebook wall instead. Then your friend has a choice about whether to take the time to pursue them.

Petitions

Again, this is your cause, and you are asking for another person's time.

"It's important," you say. "It's critical."

So many things are.

Is it dire enough for you to go to their home and have them sign it? Would you take the time to print the petition and stick it into snail mail? If you wouldn't, then it's not critical.

It is the ease of passing it on that is seducing you into action. Just because it seems easy to garner support through email forwarding, that doesn't mean you should do it.

Only send petitions if the other person has given you permission.

RULE OF THUMB

For Causes, Petitions, Requests, Sales, Promotions

It should be more work for you than for your recipient.

The ease of email has allowed us to joyfully and impulsively punch a button and feel as though we are helping the world, but it's not helpful if we are now creating more effort and less peace of mind for the person at the other end.

Requests for Donations

Yet again, this is your cause, and now you are asking for both time and money.

Would you go to this person, cup in hand, and ask them in person to shove a dollar into it? If you wouldn't, don't do so online.

If you would, and the person lives too far away to visit, send the request by snail mail.

Too much trouble? Then don't do it.

Sales Offers

You want to sell your fly-fishing collection? Use eBay or Craig's List. Only send out email fliers to members of your fly-fishing club, and only if that's a common practice in your club.

Having a garage sale? Set up a web page with a list of items and prices, then post your sale info and link on your social networking page.

Sign-Up-a-Friend-and-Get-It-Free Promotions

Before you do this, *always* check with your friends to see if they are interested in this product or event. Only sign them up if they say yes. Otherwise, you are pimping your friends.

Risqué and X-Rated Items

Tread carefully here. Only send such items to friends you are absolutely sure will appreciate and want them.

The Impulses behind Enthusiastic Intrusion

A Way to Connect

Most of us are terribly busy. We have friends or relatives we care about, but we don't live close enough to visit readily. A forward is a way to say, "I haven't forgotten you," "I think about you from time to time," "I saw this and I thought of you," or "I still know you and what your interests are."

However, this is not real connection. You are not sharing something of yourself, except indirectly. You are sharing a political view or your sense of humor, but not personal news (e.g., that you've propagated a new tomato or had a cold for six weeks that turned out to be pneumonia). A forward isn't personal. It doesn't advance your relationship.

Far better to drop a two-line email that says:

> Up to my eyebrows in alligators, but thought of you when I watched Junior run the 100-yard dash. Remember our jerseys?
>
> "Love, Me"

Now, that's a connection. You've shared something about your life and you've referenced your relationship. Good job.

Meaning Well

You care exceedingly. You want a cure for MS. You want animals to be treated kindly. You want your friend to be saved.

In your fervor, you involve your friends, perhaps for their own good, perhaps because you know they share your concern.

The rule of thumb is this: If you both belong to a group with this cause — you both belong to the same church or religion, you both volunteer at the animal shelter — then specific forwards related to that interest might be fine. Just check first.

> RULE OF THUMB
>
> If you have permission and you both belong to a group with this cause, topical forwards are fine.

However, if you are trying to influence someone to your point of view, don't forward an email. Meet in person; talk on the phone. Communicate directly in a way that lets you exchange your views. Otherwise, you are putting yourself in a one-up position of knowing what's better for them than they know themselves. And, in putting yourself one-up, you put the other person one-down. It's a boundary violation. Whether you see it or not, you are harming your relationship.

Not Thinking or Caring about the Impact on the Other Person

Sometimes we're just so deeply into our own interests, we don't realize that not everyone is as rabid as we are. We're delighted with our discovery of the working piano made out of popsicle sticks by two ten-year-olds in Topeka. The mayor got married to the sound of Bach and the scent of Fudgsicles. Who wouldn't be interested?

Lots of people. Lots of other people are just as deeply into their own interests. They don't have time for all the wild eccentricities in the world.

Redirecting Your Enthusiasm

There are two basic ways to let others choose the degree to which they'll involve themselves in your messages.

Post It

Whatever it is you want other people to see, put it on your Facebook wall or timeline. Social media sites like Facebook take all the pressure out of forwarding. Put your causes, jokes, video links, promotions, chain letters, and garage sale notices there.

Any friends who are tracking you will eventually receive your information, and they will look in their own sweet time. You leave them the choice of how much time they want to spend and how much effort they want to take in response.

Is this too much trouble? Then those cute toad pictures are not that important after all.

Get Permission

If you feel you can't possibly survive without being able to forward, feel free to copy the following form and send it to your friends.

> Hey friend,
>
> I like to forward things from time to time, but I would like to know what your own limits are with regard to receiving forwards. Please fill this out and return it to me.
>
> ❒ Don't send me any forwards.
> ❒ Send me all your forwards.

- ❒ Only send the following types of forwards:
  - ❒ Cute puppies, cats, other animals
  - ❒ Inspiring pictures
  - ❒ G-rated jokes
  - ❒ Jokes about: ______________________________
  - ❒ Pictures
  - ❒ Video links
  - ❒ Causes I am interested in (please list some)
  - ❒ Religious topics
  - ❒ Political topics
  - ❒ Requests for donations
  - ❒ Requests for action
  - ❒ Information about things you are selling
  - ❒ Chain letters
  - ❒ Information relating to this activity that we both share:

    ____

    ____
- ❒ *Do not send me:*
  - ❒ Cute animals
  - ❒ Inspiring pictures
  - ❒ Jokes of any kind
  - ❒ Jokes about ______________________________
  - ❒ Pictures, no matter how beautiful
  - ❒ Videos or video links, unless you tell me the topic and length
  - ❒ Information about worthy causes
  - ❒ Religious topics
  - ❒ Political topics
  - ❒ Requests for donations
  - ❒ Requests for action
  - ❒ Information about things you are selling
  - ❒ Chain letters

- ❐ Information relating to this activity that we both share:
  ____
  ____
- ❐ Send me as many forwards as you want.
- ❐ Send me no more than ______________ forwards per ____________.
- ❐ *Send me no more than:*
  - ❐ 1 forward per day
  - ❐ 1 forward per week
  - ❐ 1 forward per month
  - ❐ 1 forward per year

> RULE OF THUMB
>
> Are you thinking it will take too much time to send out the above form, keep track of each person's preferences, and respect each person's limits?
>
> Then don't forward things to them.

Of course, you can use this same tool to set your own boundaries on forwards with your friends. Fill it out, and send it to them. Pay attention to who remembers and who disregards your limits. Make a stronger request if your first was disregarded.

Chapter 6

GAP VIOLATIONS

Most of the violations that occur in cyberspace are intrusion violations, but gap violations do also occur, and they often carry an ambiguity that leaves a person in suspension.

For example, Magda sent a text to her mother, Essie. It said, "I've just been in an accident."

Essie replied immediately, texting, "Where are you? Do you need help?"

There was no answer.

Essie called Magda's phone. It went straight to voice mail. She left a message. She called Magda's husband and also ended up in voice mail. She texted him, "Where is Magda? Are you with her? What happened? Who's with the children?"

Five hours passed before Magda replied with a text: "I'm in the ER waiting 4 surgery."

Essie texted back: "What ER? Why do you need surgery?"

Another hour passed before Essie found out what hospital her daughter was in and could go to her.

What happened to Essie in the six hours between the first text and the location information on her daughter? She put her life on hold. She quickly finished her shopping, refueled her car, and canceled her appointment to read to her friend at the nursing home. She told her mah-jongg club she wasn't coming and hurriedly called three friends before she found someone to replace her at the table.

Then she worried. What if her son-in-law was also in the accident? She didn't know where her grandchildren were or if they needed her. She tried to tell herself that if her daughter's family needed her help, they'd call, but she could imagine an accident that was so grave that the children would be without care.

A phone call from Magda would have allowed for quick questions and quick answers. In the same amount of time it took to text "I've just been in an accident," the following phone conversation could have taken place.

"Mom, I've just been in an accident."

"Are you hurt?"

"I think my leg is broken; they're taking me to the hospital."

"Which hospital?"

"Mercy General."

"Is Roge with you?"

"He's coming, but he's picking up the children first."

"Are the children coming to the hospital?"

"Yes."

"Do you want me to come and take them home?"

"That would be great. Thank you."

"I'll be right there."

There are some definite advantages to texting, but the quick two-way sharing of information essentials isn't one of them.

Magda could have answered her phone by saying:

"I can't talk now. They're loading me into the ambulance."

"Where are you going?"

"Mercy General."

"See you there. Bye."

But an unanswered text could mean anything. The battery died. The signal got too weak. The accident was horrendous. Magda blacked out after sending the text. The accident occurred in a terrible neighborhood and she's alone and threatened. An antisocial bystander stole her phone. The EMTs are working with her. Magda is dead.

It's this ambiguity that maroons a person.

An unanswered text is therefore a gap violation, particularly if the sender needs information in order to proceed. "Are you driving to rehearsal, or do you want me to drive?" Without an answer, the sender doesn't know if she should go back home and wait, or continue on to her friend's house and pick her up.

People can be thoughtless in any medium, but electronic messages, because of their ease for the sender, can blind us to the work we're making for the recipient.

Ambiguity and Boundaries

Ambiguity pervades much of electronic communication. Texts, tweets, posts, and emails offer infinite varieties of possible issues and pseudoissues. They might or might not be real issues, but they do have the power to be boundary violations.

Ambiguity that adds work for the recipient can be both an intrusion and a gap violation simultaneously. Phrasing that sounds angry rather than neutral, absent of a direct expression of the sender's feelings, is both. Indirect or unexplained anger is very intrusive. And the missing information, whether careless or intentional, is an important emptiness.

The recipient will then have to decide whether or not to get clarification or explanation, and will also have her own feelings to manage, due both to the ambiguity of the message and the possibility that the sender was either careless or indirect.

Written messages may seem accurate and efficient, but you can't hear inflection or tone, so it's easy to read in meanings that aren't there — and to take offense or feel slighted.

Furthermore, you can't always be sure someone has even gotten your message. Emails get lost. Text message delivery can be delayed. A text may arrive at your computer but not your phone, or vice versa. You might think you weren't responded to, while the message actually went astray (probably to the land where those missing socks go).

Ambiguity can also result from a message being sent to multiple people at one time. For example, you don't know who else is seeing a text or email sent to you. You don't know what text of yours is being forwarded by your friend or relative — or to whom.

Imagine that you email your son, and he forwards your email to his wife's Facebook address. Then she writes, "There goes your mother again!" and neglects to mark her message as private.

Now you've seen her message, which doesn't sound like a compliment. Did she post it on her wall by mistake? She (and your son) may or may not have thought it through — that you are a Friend and can see all their posts. Did she totally misunderstand what you meant in your message to your son? Do you let her know that you saw what she wrote? Do you explain it to her? Do you talk to your son about it? What do you do?

Even a pseudoissue can weaken the boundary between you and your daughter-in-law. These situations create a mess of energy-consuming (and occasionally crazy-making) concerns.

This is itself a boundary violation — a bunch of issues, some real, some not, some your business, some not your business — violating your peace of mind and calling for some action that takes time and energy from your own pursuits. And the violator is not necessarily a person, but the medium itself.

It all causes unrest, unease, and a sense of too much to do that floats within your body somewhere, producing a subliminal energy drain — and thus a broken boundary.

Chapter 7

VIOLATING THE BOUNDARY OF AN EXPERIENCE

I was at dinner in a fine restaurant. At the table next to mine, a couple was celebrating their anniversary. He was on the phone talking to his buddy about a basketball game. She was on the phone to her friend gossiping about another friend.

At a family breakfast the morning after the funeral of a well-loved patriarch, half the adult family members were fiddling with apps on their phones. Some members tried to talk as a group about their memories of their ancestor, but the conversation fizzled.

These are gap violations, creating distance where there's a possibility or a goal of intimacy or connection. At an event where two people could be fortifying their intimate bond, or a family could be sharing a life passage or coming together for healing, they were interacting with absent people or with an object.

A family dinner, hanging out with a friend, strolling with your sweetheart, a bridesmaid breakfast, a family reunion — these are all occasions of bonding, of strengthening your connection with people you care about. On these occasions, except in an emergency, put away the phone (and any other electronic interrupter). When any member of a group is in electron-land, it affects the energy of the whole gathering, particularly when tender or intimate sharing is attempted.

No matter how polite people are about it, a phone call or playing with your phone is an interruption for everyone involved. Basically, you put everyone else outside your circle while you interact with your device. You say to everyone, "Your time isn't as important as mine. Our relationship isn't as important as the person I will interrupt you to talk with (or the device I will give my attention to)." You create a distance violation, putting a gap where connection belongs.

Beyond Emily Post

I have a group of friends who gather monthly at my house. Together we watch an entire TV series, all seasons, from start to finish.

After we had been meeting for some years, iPhones were invented. Our two most tech-savvy members got them, and they'd sit together on the couch and compare apps while the rest of us watched the show. When we objected, they'd put them away. Then another friend got an iPhone, and we had to go through the objections again.

Then iPads were invented. Yep, we had to object again.

With each new innovation, we all have to feel our way into the appropriate range of behavior in each situation. We've left Emily Post far behind.

Protect Your Life Experiences

If the primary reason for the event is human connection and interaction, turn off your devices. It is also okay to ask others to do the same. (Leave some wiggle room for people who might need to make or take a genuinely urgent call.)

The same rule applies if the primary reason for the event is to enjoy entertainment together, especially if the entertainment involves listening or watching in a quiet room.

Private times with people you are close to — relationships you value — should be protected from intrusion by devices or other persons. Romantic dinners, drives, or dates; dining, sitting on the porch, or strolling with your close friends, your children, or your parents; getting together with a gang of your friends to watch old movies — these are all cherished experiences. Be fully present for them.

If you are just hanging out with someone — running errands, cleaning the basement, taking the car for an oil change — slight interruptions of texts or phone calls are not a violation of the experience, provided *you have other times where you do focus on nourishing the relationship*.

However, it's a boundary error to spend a long time on the phone so that you are more with the caller or the app than the person next to you. You're basically leaving your current companion alone, in a situation where they can't do anything else. Of course, if they have their own phone and are perfectly happy chatting with someone else as well, then why are you even together? Next time, hang out with the person you are on the phone with.

Sometimes friends get together to support each other in a healthy practice. They walk every Tuesday morning, they weed blueberries together, they go to the gym. Quick texts and phone calls aren't so interruptive in these situations. The focus is more the task than the relationship.

The following table sums up appropriate cell phone use in a variety of situations. (Note that in the far-right column "Interact with Phone" refers to using apps to work or to play solo games.)

PERSON	CONTEXT	PURPOSE OF THE EVENT	TEXT	RECEIVE PHONE CALLS	CHAT ON PHONE	INTERACT WITH PHONE
Date	Private or semi-private	Developing a relationship	No	No, unless it's the "emergency" call to get out of a first date	No	No
Intimate partner or spouse	Private	Connection	No	No	No	No
Intimate partner or spouse	Semi-private	Connection	No	No	No	No
Intimate partner or spouse	Public	Hanging out	Quick ones only		No	No
Intimate partner or spouse	Public	Errands	Yes	Yes	Brief chat, if okay with the other person	If you have other bonding times
Your child or your parent	Private to semi-private	Connection	No	No	No	No
Your child or your parent	Public	Connection	No	Quick ones only	No	No
Your child or your parent	Public	Errands	Quick ones only		Brief chat, if okay with the other person	No
Close friend	Time alone, possibly semi-private	Connection	No	No	No	No
Friends	Group	Connection and fun	No	No	No	No
Friends	Routine	Health and exercise	Quick ones only		No	Yes, if the other people are similarly occupied and you have other occasions together
Friends or family	Emergency	Handling it	If relevant to the situation			Yes, if waiting long hours

Exceptions

Let's say you've trained your children to text you when they reach school or get home from school or an event. This is a brief interruption. Just let the other person know in advance, and turn off the phone afterward.

Here are examples of appropriate exceptions:

- A loved one is in an unusual situation that you want to monitor.
- You're "on call" while your child takes an important test.
- Your mom is getting test results from a doctor.
- Your husband has an important interview.
- Your daughter could go into labor; you might even need to talk a bit when she calls or head for a hospital.
- You're waiting for important news.
- The job offer may come through.
- An appointment must be confirmed.
- You might get a test result.

> **RULE OF THUMB**
>
> Let your companion know that you have to keep your phone on and that you may be receiving a communication.

Simply let the other person know that this could be happening, to be respectful of the relationship and the other person's time. *We do so much better when we know what to expect*. Otherwise, the other person settles into the event thinking it's your special time together and then gets jolted by the intrusion.

Venue-Created Boundaries

A venue may, in itself, call forth certain limits in order to protect the experience of the participants. A ringing phone while you are meditating or praying in a sanctuary intrudes on everyone's inner experience and spiritual process.

At a conference, in a movie, in a classroom, at a library, in a common work space — any place that requires focus and concentration — setting limits protects the experience of all the participants.

This table sums up appropriate venue-related boundaries:

VENUE	SEND TEXTS	RECEIVE TEXTS	ANSWER PHONE	CHAT ON PHONE	EMAIL OR SURF	USE APPS
Theater	No	E	E*	No	No	No
Religious event	No	E	E*	No	No	No
Classroom	No	No	No	No	Only if class-related	
Conference	No	E	E*	No	Only if conference-related	
Common work space	No	Yes, if work-related				

E = In true emergencies only (set your phone to vibrate)
*** = Whisper, "Hold on a minute," then leave the room before talking.**

Loud, Chaotic Gatherings

Rock concerts, sports events (except golf or tennis, which are quiet venues during play), company picnics, school rallies, and huge family reunions can have more flexible boundaries. A ringing phone will probably not bother anyone, and if you drop out of the action to text or chat now and then, you may not be missed.

However, if you are there with (or for) another person, and the goal is to experience this event together, then *the relationship is still the primary reason for being there, despite the presence of all those other people*. Keep your calls and texts to a minimum, and put your phone away.

School

It's tempting, especially if you have a boring teacher, to mentally check out and text or surf to entertain yourself. It's hard to resist this impulse, because rebelling is built into us at adolescence. However, what you do at each level of schooling really does make a difference for your life.

Being present in class gives you a shot at absorbing important information, even if the teacher could moonlight as a sleeping pill. So don't text, email, or play with apps in class, and only surf if doing so is part of the class. Otherwise, you violate your own boundary — that of showing up for your life experiences.

Although this is not posted on the front of the building, school offers two other opportunities beyond an education: to learn to operate in the culture, and to practice and develop social skills.

Learning to operate in the culture involves knowing when it is appropriate to use digital devices and when it isn't. In class, in a conference with a teacher, in a work-study group or lab with other students, or at a school club, it is not appropriate to use your phone unless it's a resource for a project. By showing up for these interactive encounters, you get to find out what works and what doesn't in accomplishing a task with others, which will give you important social skills that will be valuable to you later.

You might also upset the person or people you're ignoring. A critical social skill is knowing when another will be disturbed by your behavior and learning to discriminate when you should or shouldn't adjust your actions due to their consequences to others.

You'll miss these learning opportunities if you're texting or playing with apps.

On the Phone

When you are on the phone with someone you care about, don't be surfing, gaming, emailing, or texting at the same time. It violates the experience of interacting with that person. We are all savvy to this now: the sound of clicking in the background, pauses that are too long, a person's tone being partially absent. If you want a relationship with that person, show up for it, including on the phone.

And when you are talking with someone on the phone and another call comes in, don't take it unless it's genuinely important.

Driving a Car

Don't text or use your phone while you drive. It makes you dangerous — seriously dangerous. Texting or calling while driving kills thousands of people every year.

I can tell when the person driving the car in front of me starts using a phone; can't you? Their car wobbles. It starts weaving, sometimes quite a lot. They slow down. There's no doubt that their attention is divided. It only takes a split second for that wobble to go too far and cause a disaster.

A hands-free phone — *if you don't have to punch any buttons* — is permitted in some areas and states. But if you aren't operating the controls by voice command, if you have to look at something to find the number, call, see the message, change the volume, or hang up, *don't* — not while driving. (And even when you are talking hands-free, 37 percent of your brain is now unavailable for driving.)

Therefore, don't text or use your phone while you drive.

These same rules apply if you're biking, skateboarding, roller skating, sailing, flying a plane, or riding a Segway. Fast transportation requires your attention.

CONTEXT	SEND TEXTS	RECEIVE TEXTS	ANSWER PHONE	CHAT ON PHONE	EMAIL OR SURF	USE APPS
School	No	No	No	No	Only if part of class	
Work	Only if work-related		Only if work-related or emergency	Only if work-related		
Home, your time	Yes	Yes	Yes	Yes	Yes	Yes
Home, family time	No	No	Maybe; determine this in advance with your family	No	No	No
On phone	No	No	Emergencies only	Not to someone else	No	No
Driving	No	No	No	No	No	No

When Others Violate Your Experiential Boundaries

One of the most common boundary violations occurs when people disturb others in a normally quiet public place by talking on their phones. How should you respond to this type of violation?

In a place where you know the people involved, such as your work space or church, a simple, direct statement usually works:

"Would you please turn off your phone before services begin?"

"Would you mind setting your phone to vibrate? I'm concentrating on this report for our group and it's complicated, so I need to focus."

Strangers can be wild cards, however. So if a stranger intrudes by talking on the phone — at a movie, in a library, or at a conference — it's safer to ask the manager to handle the situation. Or simply move to another section.

With a friend who is always connected, don't wait for the first inevitable boundary violation to occur. Be proactive. Say, "Would you being willing to disconnect, just for the time we're having lunch? I'd like just to be with you."

When you're talking with a friend who tends to play with his computer or e-tablet even while he's on the phone, say something like, "Jay, I can tell you use your computer while we're talking on the phone. Would you be willing to give your full attention to our conversation?"

If you're riding in a car while the driver texts or talks on the phone, don't be silent. Speak up: "Elle, I appreciate you driving us to the workshop, but I get concerned whenever you pick up your phone. I've learned how dangerous that is and how quickly a tragedy can happen. Would you be willing to turn off your phone until we get there? Or if someone calls, I could answer it and tell them you'll call back."

If, in response to one of these requests, the other person gets angry, takes offense, or refuses, it's best to back off. You've now learned several important things: 1) your friend doesn't care about the effect he or she has on you; 2) your friend's behavior is not likely to change in the future; and 3) there's a limit to how close your relationship can be. (You've also learned to not ride with that friend in the future!)

Pavlov's Ringtone

I'm harping on this, I know, but look at the numbers. Despite nationwide campaigns, state laws against it, and the risk of a sizable ticket, 70 to 84 percent of people answer the phone while driving, with the higher percentage among those between the ages of twenty-one and forty-four. Within this group, a higher percentage of answering drivers are men, who do not have centuries of multitasking folds in their brains. Besides, driving is itself a multitasking activity (watching road, monitoring mirrors, traffic, and speed, adjusting windshield wipers, dimming lights, etc.).

The numbers:

- With respect to traffic safety, drunk drivers and drivers using cell phones are equally impaired, although the mechanisms are different. There isn't much improvement using hands-free cell phones versus handheld, because the conversations themselves are distracting.
- There's a 37 percent decrease in brain availability for the task of driving while listening to another speak on the phone, even when the driver is not holding or dialing a phone. It's somehow more captivating and narrowing to listen to someone on a phone than to listen to someone in the car.
- In comparing the ratio between doing various tasks while driving and the amount of time the driver's eyes were off the road, one study found that texting yielded the highest ratio, increasing the odds of an event (problem while driving).

"Put simply, tasks that draw drivers' eyes away from the forward roadway were those with high odds ratios. For example, texting, which had the highest odds ratio of 23.2, also had the longest duration of eyes off forward roadway (4.6 s over a 6-s interval). This equates to a driver traveling the length of a football field, at 55 mi/h, without looking at the roadway."

- The death rate for drivers between the ages of twenty-five and forty more than doubles from crashes due to distraction from a handheld phone.

So why, with so many warnings and hundreds of research studies backing them up, are we still answering the phone while driving?

Subtract six from your age. That number represents the approximate number of years you've been answering the phone. You've been solidly trained to turn your attention to a phone, whether it rings, chirps, whistles, or plays Beethoven's Fifth. Pavlov told us about this. Reinforced conditioning creates automatic behavior, and automatic patterns overrule cognitive decisions because we react faster than we think.

This conditioning, for most of us, is deeper and more embedded than the more recent development of a traveling phone and dire statements from accident experts.

To be truly safe, turn off your phone when you get into the car, or turn off both the ringer and the vibrate function. Put your phone or purse in an odd position, or twist your watch to the other side of your wrist to remind yourself to turn your phone back on when you reach your destination.

The Ultimate Violation

I was returning home on a country road, driving just beneath the speed limit. The lowering sun had turned the world golden. Though the sun was behind me, the light was so sharp that tree-shadowed parts of the road were black. I wasn't worried; I was nearly the only car on the road.

As I slowed for a curve, a bicycle burst from a side road that had been obscured by darkness. I only saw the bike at the last moment. I swerved instantly, and the bike and rider went on their way, probably with heart in throat. I know mine was.

And then I realized that if I had been lured by the quiet of the evening and the scarcity of traffic into thinking it would be safe to text or talk, I would never have seen that boy in time. I needed that split second of awareness to save his life (and my future peace of mind).

If ever I needed a lesson about how critical it is to resist the temptation to text or phone while driving, no matter how safe it appears to be, that moment delivered it. I hope you'll give yourself a moment to consider how it would have felt to watch your car take a boy's life, had you been the driver with a phone in your hand.

Chapter 8

HANDLING BOUNDARY VIOLATIONS

Diagnosing a Violation

When your own boundary is crossed, red lights don't necessarily flash. You don't always have a clear idea of what the problem is. Your first hint may be a feeling of unease, a subtle sense of something being off, a feeling of being diverted, a sudden loss of energy, or an unexplained sense of shock.

You may get very busy, start fixing or explaining things, reach for a candy bar, get defensive, feel like retaliating — before you can even articulate what bothered you. This is called automatic behavior. It is either self-soothing or self-protective, an ingrained coping mechanism.

When you catch any such signal that your boundary may have been crossed, you can use the following protocol to bring the bits of the boundary issue into clarity.

1. Pause.
   - Backtrack to the moment when your energy changed.
   - What happened right before the change?
2. Who did what?
   - Was it a comment, a word, a silence, a tone, an action, or a lack of action?
   - Think back over the triggering incident and relive it, moment by moment.
   - See if you can find the fulcrum that tipped the balance for you.
3. What do you need to do for yourself?
   - Do you need time to think about it?
   - Do you need to consult with someone?
   - Do you need to give yourself care?
   - Do you need to pause and make room for your feelings?
4. What about the other person?
   - Did he inadvertently ignite a personal issue of yours that he doesn't even know about?
   - Is she a repeat offender?
   - Was this idiosyncratic, or do they do this all the time?
5. What would be the effect on you if you let it go?
   - What would it cost you if this keeps happening?
   - How much work would you have to do to keep handling the effects of it?
   - How would that affect the relationship?
6. Would you be safer or your energy more protected if there were a boundary?
   - Would a boundary preserve something important in the relationship?
   - When you think of having a boundary protecting your energy, how does that feel?

- What happens to your energy when your imagined boundary is in place?

7. What boundary would you like to have? Jot down a quick description.

   ____

   ____

   ____

   ____

   Notice how it feels, imagining the existence of this boundary. Notice how it feels in your body. Notice how it feels in your mind.

   If the thought of a reinforced boundary makes a positive difference, use the exercise on Boundary Building in chapter 3 (see page 18), starting with step 4.

Yikes! You Mean I Have to Set Each of My Boundaries Myself?

Yes. Remember, it's your job to protect your life and your energy. No one else can (or will) do it for you.

As with any new skill, boundary setting can feel awkward at first. Many of us fear we'll lose a friend if we set a boundary. But think about the cost to the friendship if you *don't* set a boundary. Over time, the friendship is likely to fade anyway if your boundaries continue to be trampled (or if you keep trampling on someone else's).

In the short run, setting boundaries may take extra energy, but in the long run, you'll have more vitality, more power, and greater effectiveness. You'll also have cleaner relationships and greater intimacy with the people you care about. It's a worthwhile investment.

PART III

PROTECTING YOURSELF AND OTHERS

Chapter 9

FOCUS BOUNDARIES

Intrusions on a Corporate Scale

When you move into a new abode, your rooms are filled with boxes. Your new home is cluttered. Then you unpack, your things find their right places, and you recycle the boxes. Bit by bit you enjoy the receding tide of clutter. As harmonious space emerges, you can feel the difference in your body.

Through the multiple enticements of the Internet, we have the capacity to get a superficial awareness of many things that are going on in our world. Issues, scandals, short-term marriages, minor celebrity scuffles, injustices, and political brouhahas are all splattered about in cyberspace.

Just as a cluttered room is chaotic and can drain your energy, filling your mind with the diversions of the Internet can also cause a type of clutter — one that can take you away from the very important business of living your life.

You Can't Tweet Just One

For years, I specialized in food addiction, researching the causes and promoting recovery. At a professional conference one year, I was shocked to discover that a major food producer actually employed three scientists to study how to increase the addictiveness of the foods it manufactured. The lack of ethical boundaries and the premeditated harm to the public appalled me.

Web browsing, tweet reading, online gaming, and YouTube watching can be like an addiction. You can forget your troubles, be entertained, put your brain to sleep. They can call to you while you're doing legitimate computer work. Some folks actually do get addicted to some aspect of cyberspace, suffering the same consequences as any other addict: loss of health, friends, money, and employment.

Your own friends can be innocent pushers. They send you a link in an email that says, "I was moved to tears." You want to know what mattered to your friend; you want to be moved, too. So you click the link and watch the touching video. Arrayed next to it are ten other poignant videos. You wander through a couple of those. You are invited to Like them. Liking takes you to Facebook, where you can wander some more. Then you go to Twitter, send one tweet, and read a few others. You become aware of lots of things going on with people you know, and many people you don't.

You are invited to become a voyeur.

Meanwhile, an hour of your life has vanished.

How many times have you started your day with a clear plan, only to have your agenda smashed to pieces by emails, texts, and instant messages? An hour on the computer turns into three hours. The cat is lonely, you've skipped breakfast, and the project you were going to start is a distant memory. Now your day is in tatters, you feel scattered, and your energy is low.

Whose Agenda?

The problem isn't just that this infinite arcade exists and waits for us to wander through it. It's that Internet enterprises are aggressive in their efforts to get your attention. If you haven't responded to a Friend bid on Facebook, you get a reminder. LinkedIn will let you know how many people are languishing because you haven't kept up with their invitations. You'll be twigged that you have a new tweet follower. Everywhere you go, you are encouraged to Like something else.

These practices create a never-ending cascade of guilt-inducing intrusions. It's as if it were raining clutter. Added to the legitimate items on your to-do list, you can feel as if you are living in constant catch-up mode.

What to Do?

Any time you are presented with a choice or opportunity in cyberspace, ask yourself this simple question: *Whose agenda does this serve?*

Imagine that Facebook has reminded you that you haven't added Karon to your Friend list. Whose agenda is this? Who is being served?

Is it Karon? Is it you? Or is it Facebook?

The Internet social media providers are giants. How did they get there? They found ways to get visitors to their sites. Do they really care about your relationship with Karon? No.

Does Karon really need you to click her into your Facebook list, or does she already know you care about her? If you can't get to her request for a month, will it make any difference to your friendship?

How will Friending Karon on Facebook benefit — or hurt — you? How will Friending Karon *right now* benefit — or hurt — you?

By taking authority back from Facebook and getting clear about whose agenda is in operation here, you will know what to do. Depending on your answer to that question, you might delete her Friend request, wait to Friend her over the weekend, or Friend her immediately.

Want to see the pictures of your neighbor's new grandchild? Catch up with what your sister (apparently allergic to the phone) is doing? Fine, go read those Facebook walls — but at a time that works for *you*. Also, notice how many of these are one-way streets in which no personal involvement actually occurs.

The single most important principle for setting boundaries in cyberspace is this: *Get clear about your own agenda, and then follow it — not someone else's.*

Fencing Goliath: Specific Tips and Guidelines

- *Use your focus to create a boundary.* When you focus on what matters to you, you will automatically fence out superfluous subjects and activities.
- *Protect your concentration.* Open only the applications you actually need for what you are doing. Keep your email turned off until you've addressed your main concerns. Turn off your phones until you have completed your task.
- *Put boundaries around your visits to social media sites.* Schedule those visits to occur after you've finished your work and spent time with your loved ones.
- *Change your notification set-*

> **CHANGING YOUR NOTIFICATION SETTINGS**
>
> For the most up-to-date information on changing your settings, visit the Help menu at any sites you use or www.boundariesworld.com/updates, which will be updated on a yearly basis, for at least ten years following this book's initial publication.

tings on social media sites so that you aren't contacted with every little poke from every little nose.

Facebook:

- On the Facebook menu bar, click the arrow on the far right, next to *Home*.
- Choose *Account Settings*.
- On the Account Settings page, choose *Notifications*. Now you'll see a long list of categories about which Facebook will email you.
- Choose *Edit* next to Facebook. Now you've found another long list.
- Go down the list and click on or off to indicate whether you want to be emailed when a particular thing happens.
- Be sure to save your changes (in a faint box at the end of the list).

Twitter:

- Press the arrow to the far right of the Twitter menu, just past your twitter name.
- Choose *Settings*.
- On the Settings page, choose *Notifications*.
- Choose what emails you want Twitter to send you.
- Save.

LinkedIn:

- Find the little arrow to the left of your name, on the right side of the LinkedIn Menu bar.
- Choose *Settings*.
- On the Settings page, choose *Email Preferences*. Here you can make several choices about what notifications to receive and how often to receive them. A nice feature is that you can

also write a note that explains how often you check emails, so that colleagues know how soon you might receive their message.

- Press the Save button.

Well-Intended Intrusions

You're deep into your research about the tidal habits of the single-footed kelp creeper. You've just had a brilliant insight about the muscular configuration of the wee jumper that could change the life of anyone with a prosthetic limb.

You hear a ping. A message flashes onto the screen.

Meggie,

What great good fortune that you're online. I'm over the moon. I scored two tickets to the Straw Hat Hollerers Friday.

Go with?

Gina

You're wordless, and not just because you've always wanted to see a live performance of the Straw Hat Hollerers. You've been shocked out of the deep concentration that held you in the part of your brain where you access those leaps of creative discovery.

The instant message slammed you into a beta state, and you're disoriented. Finally you're able to come forth with:

Gina,

That's amazing. Of course.

Meggie

This is the point at which to set a boundary so you can get back to your brilliant discovery, but two things mitigate against having the wherewithal to do so: You're disoriented and still scrambling; it's very hard to interrupt the joyful enthusiasm of someone you care about.

> Meggie,
>
> I couldn't believe it. I was just trolling through Twitter when I happened upon this fan site for the Hollerers. It's so fun. It has a store where you can order your own monogrammed straw hat, and you choose whether you want the style Met wears, or Jet's, or Let's. And they have burlap scarves in all colors, with "Hollerers" stitched along the side. And…"

You bring back the kelp creeper's article and put the IM window to the side so you'll notice when Gina has finished sending installments of her Hollerer discovery. Now your attention is divided between the creeper article and Gina's enthusiasm. Finally, your over-the-moon friend wanes and goes away.

You try to remember your insight. It's lost.

You page through the kelp creeper website, trying to retrigger it. You reread the article you were immersed in before Gina showed up. Nada. It won't come back.

This is the problem with instant messages. They arrive when you are already on the computer. You may be working; you may be catching up on email; you may be studying or discovering something. Whatever you are doing, the IM interrupts you, smashing into view without warning. And even the briefest exchange can hijack your attention.

So don't let it happen.

You have to take responsibility for putting up blocks that prevent constant access to you. No one can do it for you.

Just Don't Open

Fortunately, you can use electronic barriers to create a boundary.

- *Don't open your Internet connection.* Turn Wi-Fi off while you're doing concentrated work.
- *Don't open your browser.* If your work does not require use of the Internet, don't open your browser until you're done or you no longer need to concentrate. Keep Safari, Firefox, and Internet Explorer turned off while you're focusing.
- *Open only the websites you need while working.* If your work does require use of the Internet, access only those websites you need. Discipline yourself to ignore links within those sites that can lead you astray. (Superfluous links can be like bread crumbs leading you down a thousand paths through a million enchanted forests.)
- *Don't open your chat or message application.* If you don't need to chat with anyone to do your work, don't open the program that makes instant messages possible.
- *Adjust your chat settings to indicate your degree of availability.* Most purveyors of instant messages give you options for controlling your visibility.
    - *Do Not Disturb*
      Clear enough.
    - *Custom Availability*
      You can chat with specific people. Only the people you authorize will see that you are online and be allowed to IM you.
    - *Invisible*
      If you have friends or colleagues who won't respect your leave-me-alone boundary, become invisible. No one will know you are on the Internet. You can still send an instant message to the person you're working with to indicate that you are actually available.

- *Don't open your email program.* If you don't need email to work, keep the program closed until you've finished the part of your day that requires your focused attention. If you need to receive emails from certain people to further your work, establish an alternative email address just for those people, and only open that account.
- *If you find yourself lured back to the Internet despite your best intentions, get the app Freedom (macfreedom.com).* It works on either a PC or a Mac, and it shuts down your access to the Internet for a specified period of time. If you want to access the Internet after setting it, you have to shut off your computer and reboot. It creates that little bit of distance that helps you stay focused, a psychological barrier — like putting money into a savings account — that puts it just enough out of reach that you are protected from your own impulses.
- *A similar app, for people who need the Internet to do their work*, is Anti-Social. It blocks the social parts of the Internet like Facebook or Twitter.

In a sense, we have to protect ourselves from ourselves. The Internet offers us unprecedented possibilities for distraction. It pulls at us. It tugs. It therefore requires us to deliberately shield ourselves from its charms. We must marshal more energy and be much more intentional about our choices if we want to keep from being pulled into its attractions.

The easiest way to do this is to not open access to the Internet in the first place. Make a rule for yourself that you'll work off-line except when you genuinely need to go online. The more your work requires you to interact via the Internet, the more intentional you'll have to be at holding a discipline and using electronic filters.

The Cost of Capricious Clicking

Never in the history of the world have we been offered such comprehensive distractions. After a day spent with some sort of screen — computer, smartphone, or tablet — many of us come home from work to sit, yet again, at a computer. Meanwhile, a spouse, kid, cat, or dog is waiting for our attention. Or they (including the cat and dog) might be busy with their own electronic devices. In an era of infinite connectivity, true connection is suffering.

What does this cost us? Time, energy, flexibility, health, connection with others, and connection with ourselves.

- *Time*. It's so easy to lose hours to a device. You can say you'll be done in an hour — and lose an evening.
- *Energy*. Do you have more or less energy when you finally leave your computer? Electronics are enervating. They drain us.
- *Flexibility*. Humans weren't designed to sit motionless for extended periods of time. Frequent movement lubricates our joints, fires tiny muscles, and requires us to adjust our balance.
- *Health*. Eyestrain, back and joint issues, carpal tunnel syndrome, and e-thrombosis are all possible consequences of

uninterrupted fixation on a screen. Sitting without a break actually increases the risk of cancer and diabetes.

- *Connection with ourselves.* While we are exploring the Internet, we aren't exploring ourselves. We become absent to our own internal processes.
- *Connection with others.* Two distracted halves don't make a whole; they make a whole lot of distance. While absorbed in electronic distraction, we are likely to be creating gap violations with our own loved ones. We can miss out on the true human connections that make life worthwhile.

Solutions

- *Create an electronics-free time of day in your household.* For an hour (or two or three) each day, turn off all computers and phones.
- *Play board games that increase knowledge or creativity*, such as Pictionary, Jeopardy, Trivial Pursuit, or Cranium. Or ride bikes or take a walk together. Play touch football. Or just talk with each other.
- *Get together in person with people you care about.* Turn off your phones. Look at each other. Talk to each other about your thoughts, your experiences, your discoveries. Share the questions you have about your life, and theirs, and life in general. Listen and pay attention.
- *Spend some cyberspace-free time alone each day.* Read. Listen to music. Write songs or poems or plays, or simply write in a journal. Paint or draw or sculpt — with actual materials. Lie on the lawn and watch the clouds drift past.
- *Protect your creative pursuits.* Turn off any device you don't need when you are in your creative mind. If you must use a computer to create — and you can afford it — get two

computers. Use one just for your creative efforts. Put on that computer only the programs that foster your creative pursuit. Simplify that computer so that you are minimizing its ability to distract.

EXERCISE: Focus Experiment

For the next five minutes, turn off your phone and your computer. Close the door.

Find something beautiful or natural to look at — a picture, a leaf, a drop of water. Put all your focus there.

Notice what it feels like to have a simple, single focus.

Slow your breathing. Allow your shoulders to drop. Relax your diaphragm. Breathe in the beauty or awareness of that object. Breathe it into your center and let your awareness sink into your central places. Feel your connection to yourself and, from that connected place, bring your attention back to the object.

Rest in that stillness, breathing in and out.

Afterward, articulate for yourself what you learned or discovered. What was satisfying about it?

If you did this experiment, you created a physical boundary by closing the door, and a boundary against interruptions by turning off your phone and computer. These boundaries protected your ability to focus. Then, as you focused, you created a powerful energetic boundary around yourself. The act of focusing *in itself* became a boundary to the outside world. You strengthened your internal power.

This demonstrates how setting external boundaries fortifies internal boundaries. Together, these can provide strong protection for you and your priorities, your sanity, your unique path in life — and your soul.

Chapter 10

PROTECTING YOUR CHILDREN

A fellow therapist and I were eating lunch together. At one point, we got onto the topic of social media. She said, "I would never put my children's pictures on my Facebook timeline. I know too much."

Granted, an occupational hazard of therapists is that when they see a child with a lone adult male, they check — not the face of the adult — but the face of the child, to read any problems there. We've heard too many true stories from clients about the predacious half brother, uncle, or friendly neighbor — and have seen too many devastating, life-altering consequences to the grown victims — to assume that every adult wants only the best for every child.

Nowadays, many of those predators troll online. With unprecedented anonymity, predators can visit chat rooms, follow tweet links, and Friend the trusting teen. You would never show a picture of your daughter to a seedy stranger on a bus. You would notice if a weird person followed you as you walked your son home from school. Yet you, or your children, can be trustingly exposing your family to an infinite number of weird strangers, who could then find you (and consequently them) easily through Google or the online white pages and, for $1.99, get full contact information.

The identifying picture on a person's wall, timeline, or web page is not necessarily a picture of the person who owns the account. Anyone can create an account, use a misleading name, install a picture of a clean-cut prosperous-looking fellow, and invent a profile — a wolf in a Brooks Brothers shirt.

This may be a worst-case scenario, but it represents a nightmare for parents and women. Above all other concerns, we want to be safe. We want to keep our loved ones, especially the little ones, safe. And protecting your children has never been more difficult. They can put themselves in harm's way, outside your awareness.

What are the odds? What are the odds your child will be a target?

- 93 percent of American children between twelve and seventeen use the Internet.
- Approximately one in five children were sexually solicited or approached online in 1999. While this percentage decreased to one in seven in a second study occurring in 2005, 75 percent of solicitors asked to meet the youth in person, 34 percent called them on the phone, and 18 percent came to their homes.

While you take a deep breath and digest this chilling information, telling yourself that you can at least take steps on your home computer, 79 percent of these solicitation attempts happened on home computers.

If you are a concerned adult, the statistics from recent studies are scary:

- One in three youths are exposed to unwanted sexual material.
- 4 percent of youths received distressing sexual solicitations that left them feeling very or extremely upset or afraid, while 9 percent of youth Internet users were very or extremely upset by distressing exposures.
- 90 percent of sexual solicitations and approaches happened to teenagers.
- 90,000 sex offenders had MySpace pages before being booted off after an independent programmer developed code to root them out. However, MySpace had to be subpoenaed by the Connecticut attorney general before revealing the exact number. Although, many people are dropping MySpace, probably because of its reputation, predators view all social media sites as a Yellow Pages of potential victims. If MySpace dies, the problem won't go away.
- "The increase in exposure to unwanted sexual material occurred despite increased use of filtering, blocking, and monitoring software in households of youth Internet users."
- A considerable percentage of teens knew they were using the Internet in a way their parents would not approve of, and most of them knew how to cover their tracks.

Although teens have become more savvy, aware that predators could lurk in chat rooms, predators have become more aggressive, sending gifts, offering solace for children in troubled homes, and even sending money or travel tickets.

While the percentage of stranger solicitations has dropped from 97 percent in the previous decade to 86 percent in 2005, teens are now also being distressed by people they knew before interacting with them online. A decrease in online civility among peers is showing up in rude sexual comments, as harassment, and in the revelation of embarrassing private information about someone.

You might say, "Things are okay at my house. My child tells me everything."

Really? Then you are beating the odds, for 73 percent of incidents were not reported to parents. Also, only one in five hundred incidents were reported to an Internet service provider, and just 3 percent were reported to authorities.

The Downside of Upgrades

Why are things so much worse today?

Kids spend a lot more time on the Internet, with almost half of today's youth using the Internet almost every day. They have more access, with 91 percent home availability. Smartphones, tablets, school and library computers, and their friends' electronic treasures mean that any kid in America can spend time online. Lightning-fast download speeds give the capacity to receive explicit pictures instantly, whereas in the olden days (ten years ago), downloading took so long that a kid with a short attention span got bored and did something else.

Video cameras are now built into most computers. It's easy, in an impulsive moment, for a teenage girl to respond to a crazy suggestion, put a body part within view of a lens, and send — before really thinking about the potential consequences.

And never forget *cui bono*: Who profits? A lot of money can be made by slipping a little program into a bundle of free games, or infesting a pop-up ad with a bug that can hijack your browser and send it to porn websites.

Just when you thought the world was getting a bit safer, with medical miracles that can shoot down microbes that once decimated whole populations, we discover these new bugs that can devour your peace of mind.

How Do You Set Boundaries against Invisible Jackals?

Make Internet boundary setting a family affair. Sit down with your computer and explore the resources found later in this chapter yourself first. Then pick the ones you think would be most attractive to your child and, together, investigate the sites. Look at videos, play interactive games, and discuss what you are both learning.

Let yourself be helped by experts who put their best thoughts into protecting children. For example, the NetSmartz Workshop has discussion starters you can use to initiate conversations with your child. Their advice: "The more often you talk to them about online safety, the easier it will get." A sample question: "How do you respond if someone bothers you while you are gaming?" See www.netsmartz.org/Gaming.

Start young. Don't wait for your children to get to a certain age before you work with them regarding Internet safety. As soon as they go online, go with them (using the resources listed later in this chapter) so that they know from the outset how to protect themselves. Be so available that they naturally turn to you if something they see frightens them.

Have occasional family discussions — say, around the dinner table — with lighthearted invitations like, "What wild things have you discovered online?" In the ensuing discussion, you can help your children learn to distinguish between what is true and what isn't and teach them to verify online information, so that they can discover that not all online sources are reliable.

"Don't Visit X-Rated Sites"

Do you think that directive would be enough to protect a child? It's not.

When interviewed, 88 percent of parents reported that they had instructed their children about cyberspace dangers and told them not to reveal personal information online. In contrast, about half of the interviewed kids reported being *told* about certain dangers by parents. You know they don't always listen, take in, or act upon what you tell them.

A list of rules isn't enough. A single video, a stern warning, bringing it up once, talking about one or two aspects of Internet risks (versus specific discussions about each type of danger) isn't sufficient. This topic must stay alive in the household.

In particular, protecting personal information is critical, not just for your child's safety but for all your children, for you, and for your entire family.

- Keep the computer in a central place so that you'll know when your kids are online and be able to keep a casual eye on what they're doing. (If you can stand it, have them keep the sound on — rather than use earbuds — so that you can hear what they are doing.)
- Help your children set privacy settings on social networking sites. If you don't know how, see chapter 12.
- Reward your child for checking with you before posting any pictures or videos online.
- Check your child's social media sites often to see what's there, and act quickly to remove any picture that reveals even a little bit about your house, your neighborhood, your vicinity, or the local school. A partial view of your license plate, a street sign, a banner in a football stadium, a school crest on a jacket, or a sweatshirt logo can make it possible for your children to be tracked. (And you would certainly not want to put any other local children at risk.)
- Make the following statement a household slogan, said so frequently that your children can finish your sentence: *Never post email addresses, personal pictures, or phone numbers*

online. (Fifty percent of predator solicitations include a request for a picture, usually a picture with sexual content.)

- Teach your children that passwords should be shared only with parents or guardians.
- Teach children to not respond to any emails requesting personal information. As a test, have a safe relative use an unknown address and a false name to send an email requesting your child's response. Reward them if they don't respond. (If they do respond, sit down and together watch some online videos about the dangers of giving out personal information, so that they can get a clearer idea of the possible consequences. You'll find those resources later in this chapter.)

SAMPLE EMAIL FROM AUNT CLEVER, ACTING ANONYMOUSLY

Hi Jeff,

You've won a free set of music videos. Just reply to this email and send us your name, address, and phone number. Be sure to tell us your top two picks, so you'll get the exact videos you want.

Also, we'll enter you into a drawing for the newest version of the Best Digital Game in the World.

We'll call the winners, so remember to include your phone number.

Good luck,

Monty
Digital Game Clearinghouse

- At the dinner table, involve the whole family in brainstorming gender-neutral screen names and email addresses. Then each family member can pick a favorite, go to the computer, and replace the old revealing screen name. Make a game out of it.

- Help your children delete emails from unknown senders.
- Point out a junk email with a link, explain why it should not be clicked on (it can harbor a bug or virus), and show how to delete it.
- Teach your children about the dangers of pop-up ads that invite signing up (and that wheedle your child into supplying names and email addresses), or of downloading free games that could have a hidden spy lurking within.

Stay in touch. Internet safety is not something you can set and forget. Stay involved in your children's Internet choices. Continue to discuss what they've learned and discovered, where they hang out, and whom they've met. Make it easy for them to tell you if something has disturbed them.

Children and teens are so accustomed to computers and life online that we grownups can easily feel that they are way beyond our own capacity with electronic media. They probably are. But we still have better judgment and a clearer idea of how a chain of events can become dangerous.

You know you'd snatch your child away from a speeding car, regardless of your ignorance of the make of the vehicle. Don't let their skill keep you from being part of their cyber-life.

Children at Risk

The children who are most at risk are those in troubled homes or whose parents aren't involved. Thus, just by being involved, you cut down on the odds against your children. If you do have a troubled home, this is one way you can lower their risk.

Working closely with your children around Internet use automatically strengthens family intimacy boundaries. They'll feel your interest and your presence. Even if you have a problem in your household, your involvement and concern will strengthen the child's sense of protection. Your relationship will be tighter. Your child will feel less isolated, and therefore less in need of the pseudorelationships that can be found online. In a world of busy adults who tend to distance themselves from current teen customs, you will have done your child the great honor of visiting youth culture.

> RULE OF THUMB
>
> If you try to visit a website mentioned in this book and discover that the URL has changed (that is, if you end up on a page that says, "link not found") go to the organization's website and enter the desired topic into its search engine. The search results should include the archived article or newer material with updated information.

Resources

- Educate yourself by using resources specifically aimed at anyone concerned about protecting children from online exploitation or harm:
    - NetSmartz 411 is a program developed by the National Center for Missing and Exploited Children. It is free. It has excellent information on all types of Internet dangers. www.netsmartz411.org
    - Check out what games your children are playing. You can research game ratings and content at the Entertainment Software Rating Board. www.esrb.org
    - Get tips for interpreting media violence to children at www.commonsensemedia.org/advice-for-parents/impact-media-violence-tips.
- Introduce your children to online resources that make it fun to learn how to deal with online demons:

  - NetSmartz Workshop has attractive, interesting, interactive, age-appropriate activities for both kids and adults. Children, families, classrooms, and law enforcement can become involved, learning how to protect themselves and others online. www.netsmartz.org/Parents
  - Girls need to have a place that is frequented and led just by other girls. For Girls, sponsored by the Girl Scouts, is an interactive resource for teen girls, dealing with bullies, social networks, sexual predators, and online etiquette. (The GSUSA changes its material and programs frequently — keeping those young minds interested — so the name of the program and content may change, but whatever replaces it will be good.) http://forgirls.girlscouts.org
- Introduce your children to safe, monitored alternatives to Facebook, YouTube, and MySpace:
  - iLand5 is a safe social networking site for kids aged five through eighteen, with many attractive activities, email, instant messaging, and games. Groups are divided by age ranges, and anyone who signs in has their identity verified. Participation is free. Go to www.iland5.com to see a demo and to download a Getting Started kit, or check out the sponsor: www.safewave.org.
- Monitor or block certain types of Internet usage:
  - Windows Vista helps you oversee the games your children play. You can decide which applications they can use and which websites they can visit. You can also set up specific times when your kids can use the computer.
  - With Windows Live Family Safety, you can specify the people your children can have contact with. (Savvy kids can get around this, so you'll still want to check on their contacts with a report that tells you what they've been doing online.)

    - For $10 a month, SocialShield finds and checks your child's friends. It alerts you if your child participates in questionable discussions about drugs, sex, or suicide. It looks for photos that identify your children. www.socialshield.com
    - GoGoStat is a free application that enables parents to monitor potentially unsafe activity on Facebook. www.gogostat.com
    - Search engines such as Google provide tools that parents can use to limit their child's search activity. Videos made by other parents, Google SafeSearch, SafeSearch Lock, YouTube Safety Mode, and the Android Rating System (which applies to apps), can be found at www.google.com/familysafety/. That same site also makes it easy to report violations on any of their products.
- Use software or applications that block, filter, or monitor:
    - PureSight has an impressive array of features — protection from cyber-bullies, monitoring of chats and Facebook, porn protection, web filtering, file-sharing controls, and reports and alerts. You can access the reports no matter where you are, and the software is tamper resistant. You can install it on all your computers for $10 or less per month (only on a PC). www.puresight.com
    - Action Alert (free) allows you to read both sides of a chat or email conversation, filters content, gives you a copy of user names and passwords, records hours of all PC activity, lets you set times when the computer can be used, sends you alerts, blocks inappropriate sites, and even lets you shut down your computer from your phone. www.actionalert.com
    - Parental Software offers a variety of programs for monitoring and controlling Internet usage, as well as a

program that lets you monitor phone calls and text messages. Programs are for the Mac and PC, and they vary in price. www.parentalsoftware.org/

- Find a comprehensive directory of online safety resources at www.connectsafely.org/Directories/Internet-safety-resources.html, including movies, posters, school and teacher aids, websites, curricula, and games.

Remember, though, that some of the above programs won't follow your child's Internet usage at someone else's house, and that there's no substitute for your personal interaction, sitting side by side with your child, at the computer.

How to Report a Predator

Amazingly, only 3 percent of Internet sexploitation of kids gets reported to officials. It's a hydra, no question, but the overwhelming number of predators is no reason to be passive. We have to start somewhere and work together to expose the bad guys.

Participate in patrolling the Internet. If you discover an email address, screen name, explicit photo, or solicitation attempt as you are monitoring your children, pass on the information to the CyberTipline. It'll take only five minutes to fill out the form. Call 800-843-5678 or go to www.missingkids.com and click *CyberTipline*.

Leadership from the Girl Scouts of America

The Girl Scouts offer their members an Internet safety pledge. For this reason alone, joining this fine organization would be a big step toward safety for your female children. (There are many other good reasons to join, including the ongoing training in leadership skills, independent thinking, taking responsibility, and developing good judgment. Most importantly, children who feel that they belong to a wholesome group are far less likely to experiment with drugs.)

Here is an excerpt from the pledge:

Girl Scout Internet Safety Pledge

- I will not give out personal information such as my address, telephone number(s), parent's or guardian's work address/telephone number(s), and the name and location of my school without the permission of my parent or guardian.
- I will tell an adult right away if I come across or receive any information that makes me feel uncomfortable.
- I will always follow the rules of Internet sites, including those rules that are based on age of use, parental approval and knowledge, and public laws.
- I will never agree to get together with someone I "meet" online without first checking with my parents or guardians. If my parents or guardians agree to a meeting, I will arrange it in a public place and bring a parent or guardian along.
- I will never send a person my picture or anything else without first checking with my parent or guardian.
- I will talk with my parent or guardian so that we can set up rules for going online. We will decide on the time of day that I can be online, the length of time that I can be online, and appropriate areas for me to visit. I will not access other areas or break these rules without their permission.

Chapter 11

VIOLENT GAMING: REHEARSAL FOR SCHOOL SHOOTINGS?

According to the American Academy of Pediatrics: "Exposure to violence in media, including television, movies, music, and video games [all of which are online], represents a significant risk to the health of children and adolescents. Extensive research evidence indicates that media violence can contribute to aggressive behavior, desensitization to violence, nightmares, and fear of being harmed."

A National Television Violence study evaluated ten thousand hours of programming, finding that 61 percent of the programs displayed violent acts. "The highest proportion of violence was found in children's shows. Of all animated feature films produced in the United States between 1937 and 1999, 100 percent portrayed violence, and the amount of violence with intent to injure has increased through the years." Also, nearly 50 percent of the films "showed at least one character celebrating the violence by cheering or laughing."

"More than 80 percent of the violence portrayed in contemporary music videos is perpetrated by attractive protagonists against a disproportionate number of women and blacks." Through such exposure, children learn that violence is a solution to problems and a way to get what they want.

The context in which violence is displayed makes a difference. When violence is shown without portraying the human cost in suffering, loss, sadness, and long-term consequences, children are not learning about violence. They are learning to *be* violent. Contrast *Saving Private Ryan* or *Law & Order UK*, which show the painful cost of violence, with *Teen Titans* on the Cartoon Network.

At what age do you think your children will be ready to view repeated episodes of fantasy violence? Eight, ten, twelve years of age? According to Cartoon Network, fantasy violence is acceptable for kids over ten. Do you agree?

I strongly urge you to go to Cartoon Network (www.cartoonnetwork.com) yourself and investigate the offerings there. Not all the offerings are questionable. Many look like fun. And some are spectacular, such as the *Stop Bullying: Speak Up* video, introduced by President Obama and narrated by child actors and popular sports figures like Venus Williams and Hope Solo. But if you get a warm, cozy feeling when your children are on Cartoon Network, you haven't investigated deeply enough.

Here's a quick exercise for you.

EXERCISE: Study Guide — Cartoon Network

1. Go to cartoonnetwork.com and click on *Shows* (menu at the top of the home page). As you scroll through the Shows homepage, how many representations of violence do you see? Include persons or animals in

assault positions, scary or rageful faces or words, and weapons.

2. Scan some of the videos that are a part of a saga. Do you see any symbolism that parallels your religious beliefs, but with a very dark spin?
3. Time yourself. Randomly click on a couple of stories or games. How many minutes pass before violent, angry, retaliatory, or frightening actions or words occur?

Here are some scenes I found as I explored the Cartoon Network website (keep in mind that my referencing these examples may cause their removal from the site, so by the time you read this, they may no longer be there):

- A teen in a school corridor, shooting air guns. (*Level Up*)
- A camera panning down the front of a slave girl's skin-tight attire and freezing momentarily on her breasts. When she thinks she's going to be given to the slave trader, she kills herself. Is the message that suicide is the way to handle being a victim? ("Slave Trader," *Star Wars*)
- A cartoon with a tux-clad rabbit playing a grand piano onstage. When an audience member coughs, he pulls a gun and shoots him.
- In friendly-sounding *Scooby-Doo Mystery Incorporated*, a bad guy trying to extort what he wants by threatening to torture someone's pet. ("Quest X Power Source")
- An airplane full of passengers with a large missile headed toward it. (*Teen Titans*)
- Adventure cartoons with unrealistically buxom women in brief, form-fitting costumes. When sexual titillation is added, violence and pleasure become associated.

You've probably noticed that, in recent years, teen girls wear tight, form-fitting clothing. Modeling revealing clothing that teen girls are prone to copying is another boundary violation — one that can expose them to additional risk at an age when they are too young to have taken the measure of the world and of what some people are capable of thinking and doing. According to the American Academy of Pediatrics,

> Children are influenced by media; they learn by observing, imitating, and making behaviors their own. Aggressive attitudes and behaviors are learned by imitating observed models. Research has shown that the strongest single correlate with violent behavior is previous exposure to violence. Because children younger than eight years cannot discriminate between fantasy and reality, they are uniquely vulnerable to learning and adopting as reality the circumstances, attitudes, and behaviors portrayed by entertainment media.

Sit with your child as he or she watches a cartoon. Notice when your child laughs. Do violent animations make your child giggle? (Don't stop this expression; just take note.) Talk about it afterward without making your child wrong. Ask what effect your child thought it had, for example, for an audience member to be shot for coughing?

Kids of all ages are becoming desensitized to another person's pain. Families will watch TV as they eat dinner, gnawing on pizza while someone on-screen is being killed. Even in cartoons, it's commonplace for the cute character to hurt someone — or to be blown away.

Considering that kids spend forty-five hours a week involved with some type of media (more than they spend interacting with parents or family), that they are learning the parameters of acceptable behavior from what they watch, and that teens who watch more than one hour of TV a day are four times more likely to commit aggressive acts when they become adults, it's even clearer that parental involvement and boundary setting are imperative.

Video Games and Online Gaming

In an opening screen for an online game on the Cartoon Network, two small animal-like characters are surrounded by scary zombies. The goal of the game is to beat up zombies using a stick. When I found this game at 1:30 PM on a Wednesday — a school day — there were 1,383 players.

The game lists the high scorers for the week. The winner killed over five and a half million zombies. Do you want your child to sit in front of a screen and manipulate an animated character so that it beats up millions of other characters with a stick?

Electronic games are consuming more time each year. Ninety-seven percent of children now play them. Half of these games are violent. Young people who play lots of violent games behave more violently than those who do not. Such gaming causes:

- An increase in aggressive thinking
- An increase in retaliatory aggression
- A decrease in helping behaviors

As many of us, both adult and child, make the transition from computer and video gaming to gaming through apps, which also offer an array of violent games, we are offered the possibility of playing games every spare moment — on the bus, at recess, at a coffee break, before the movie starts.

Many of us are repeatedly shocked and horrified as yet another kid uses an assault weapon to mow down another group of innocent children and teachers. Perhaps the American Academy of Pediatrics and Craig Anderson, in a congressional hearing, have the answer to the question "Why?"

> *Reason 1.* Identification with the aggressor increases imitation of the aggressor.... In most violent video games, the player must identify with one violent character....
>
> *Reason 2.* Active participation increases learning....
>
> *Reason 3.* Rehearsing an entire behavioral sequence is more effective than rehearsing only a part of it.... [The child] in essence rehearses this choice process.... Indeed, in many video games the player physically enacts the same behaviors in the game that would be required to enact it in the real world. Some games involve shooting a realistic electronic gun, for instance. Some virtual reality games involve the participant throwing punches, ducking, and so on. As the computer revolution continues, the "realism" of the video game environment will increase dramatically.
>
> *Reason 4.* Repetition increases learning. The addictive nature of video games means that their lessons will be taught repeatedly. This is largely a function of the reinforcing properties of the games, including the active and changing images, the accompanying sounds, and the actual awarding of points or extra lives or special effects when a certain level of performance is reached.

Children learn by observing and trying out "behavioral scripts." Repeated exposure to violent behavioral scripts can lead to increased feelings of hostility, expectations that others will behave aggressively, desensitization to the pain of others, and increased likelihood of interacting and responding to others with violence.

Rather than letting them observe part of a violent interaction, video games allow the player to *rehearse an entire behavioral script*, from provocation, to choosing to respond violently, to resolution of the conflict. Moreover, video games have been found to be addictive; children and adolescents want to play them for long periods of time to improve their scores and advance to higher levels. Repetition increases their effect.

Interpersonal violence, as victim or as perpetrator, is now a more prevalent health risk than infectious disease, cancer, or congenital disorders for children, adolescents, and young adults. "Playing violent video games has been found to account for a 13 percent to 22 percent increase in adolescents' violent behavior; by comparison, smoking tobacco accounts for 14 percent of the increase in lung cancer."

Increased skill in the special-effects department exacerbates the problem. The more realistic the portrayal of violence, the more likely it is that violence will be tolerated and learned. And on the flip side, between the imbalanced reporting on news media and the prevalence of violence in entertainment media, children are given the perspective that it is a mean world.

Operating Principles for Life

Early in life, children can make an unconscious choice about which side they want to take — victim or perpetrator.

We don't always realize that we create policies for ourselves when we are very young. Based on what we see around us, we make early decisions: "I'm going to be the one who helps." "I'd rather hurt than be hurt." "I would never do to someone else what was done to me."

These decisions then become underground pointers that determine complex and life-defining choices. In nearly every case with my clients, they'd forgotten that they'd made such a decision (until exposed by therapy), but subsequent events showed clearly that the disremembered decision was powerful and in almost constant operation.

Fear of being a victim can create armed children. Though the story is fictional, a brilliant study of this can be found in *Nineteen*

Minutes by Jodi Picoult, which I would like to make required reading for every parent and educator. (An excerpt ends the chapter.) This beautifully written, sensitive book shows the cost of child violence to all parties. It also demonstrates a very subtle boundary that parents routinely violate; it can be the final straw that breaks a child's moral back and permanently harms the parent-child relationship.

This boundary is attunement. Attunement means dropping momentarily into the world of another person and discerning their focus, before bringing your own consciousness into it. (I found it interesting that the word *attunement* wasn't even in my word processor's dictionary. Is it in your computer's dictionary? Is it in your lexicon?)

One Thanksgiving, my roommate's son, who owned a window-washing business, came to our annual celebration of gratitude. He came prepared to wash our windows — an early Christmas present. We had high clerestory windows above the living room, so he went up on a very extended ladder, his arms outstretched to reach those high windows. Meanwhile, my roommate, enthusiastically enjoying his presence, leaned out from a second-story deck to talk to him.

My roommate and her son both communicate with large, Mediterranean motions. The rest of us watched in horror as she began using those wide gestures, and then as he responded in kind, hanging half off the ladder, waving one arm wildly. I couldn't stand it and walked out to tell her to wait until he was down at least one floor before continuing her story.

It's lovely that she was so delighted to see him, but this is an example of lack of attunement. Her enthusiasm overrode her ability to see that letting it spill over in that moment could actually endanger him.

Usually a lack of attunement is more subtle. Bursting in with news of the first daffodil while someone is deep in a creative cloud, plunging immediately into what you have to say to a person before checking to see if they can be truly receptive, talking for thirty minutes before you have any idea what the other person's needs are, making non-critical decisions for everyone without discovering their preferences — these are all examples of being so deeply into your own world that you aren't paying attention to the climate of the other person.

When a child tugs on your hem and you say, "Later"; when you always share your reaction before expressing curiosity about your child's reaction; when your birthday gift is more about your pleasure or needs than your child's interests and skills — this is a lack of attunement.

Here's an excerpt from Jodi Picoult's book *Nineteen Minutes*, showing the early seeds sown by non-attunement. Peter, a kindergartner, is bullied from the first moment he steps on the school bus. This takes the form, in part, of his lunch box being thrown out the window. (Joey is his older brother.)

> "Peter," his mother sighed, "how could you possibly lose it again?" She skirted around his father, who was pouring himself a cup of coffee, and fished through the dark bowels of the pantry for a brown paper lunch sack.
>
> Peter hated those sacks. The banana never could quite fit in, and the sandwich *always* got crushed. But what else was he supposed to do?
>
> "What did he lose?" his father asked.
>
> "His lunch box. For the third time this month." His mother began to fill the brown bag — fruit and juice pack on the bottom, sandwich floating on top. She glanced at Peter, who was not eating his breakfast, but vivisecting his paper napkin with a knife. He had, so far, made the letters *H* and *T*. "If you procrastinate, you're going to miss the bus."
>
> "You've got to start being more responsible," his father said.
>
> When his father spoke, Peter pictured the words like

smoke. They clouded up the room for moment, but before you knew it, they'd be gone.

"For God's sake, Lewis, he's five."

"I don't remember Joey losing his lunch box three times during the first month of school."...

"I'm not Joey," he [Peter] said, and even though nobody answered, he could hear the reply: *We know*.

Attunement from parent to child builds a bond that gives a child trust that he or she is seen. This is fortifying in a thousand ways and gives your child a bullet-proof vest when approaching the Internet. A child who can trust that a parent will listen, who knows that a parent will take him or her in, is a child who will be able to turn to that parent when something disturbing happens online.

Contrast that trust with a dad who flies off the handle before he's even heard the whole story.

"Dad, this video popped up..."

"What were you doing?"

"I wasn't doing anything. I just thought this game looked interesting and all of sudden there were these..."

"You must have been doing something. What have you been looking at online?"

"Nothing. I just wanted to find that ranger game and..."

"You've been checking out sites that I told you not to go to, haven't you?"

"Dad, I haven't. I was just..."

"That's it! You're off the computer for a week."

Will this child continue to turn to his dad with future upsets?

Attunement builds a foundation. It's a perspective that holds an awareness of the other person's reality alongside your own. It lets you come into the space of another with respect. It's a powerful boundary builder, both for the child and for your mutual relationship.

Chapter 12

YOUR PERSONAL INFORMATION BOUNDARIES

We know who you are and we know where you live.

Sound threatening? The eye in the sky knows much more than that. It knows your web-surfing habits, your preferences, your purchases, your website visits, the emails you've sent, which of certain emails you've opened and read, and your Internet Protocol (IP) address (your computer's unique identifying number).

Dr. Latanya Sweeney, a computer scientist at Harvard, can find your name using nothing more than your birth date and your zip code. If you are younger than thirty and tell her where you were born, she can predict eight (or even nine) of the nine digits of your social security number.

PII stands for *personally identifiable information*. These are the bits of data about you that can reveal your identity to someone else.

A single piece of data, such as your Social Security number, is yours alone — and it's all anyone needs to identify or find you. Other pieces of data, which seem harmless by themselves, can be combined to find you.

Personally identifiable information:

- Social Security number
- Date of birth
- Place of birth
- Mother's maiden name
- Biometric information
  - Fingerprints
  - Retina scan
  - Face recognition from photographs (used routinely by Facebook)
  - Signature
  - Voice analysis
- Medical information
- Educational information
- Employment information
- Telephone numbers
- IP addresses (every computer has an Internet Protocol address that is recognized by other computers)
- Driver's license number
- Bank account numbers
- Credit and debit card numbers
- Work or home addresses
- Photographs

Every year, I go to both the state fair and a big flower and garden show. The past couple of years, a new item appeared at both venues. As I entered, I was handed a card. All I had to do was insert the card into a machine, fill out a form, and receive all kinds of free stuff.

I started the process. It asked for lots of information — address, email address, yearly income, and other demographics — and in return, I'd get a free pair of garden gloves, a ruler, a cheaply made stem clipper, and the right to vote on the best flower of the show.

What was really going on? Data brokers were collecting that information, which they could then sell to people-search companies, businesses, anyone.

Any information you voluntarily enter into a data-collection tool — such as the form you fill out to enter the car-giveaway sweepstakes at the mall — becomes the property of the vendor and can eventually enter the public domain.

It astounded me that people were lining up to use these machines. I'll bet not one person there actually needed another pair of gardening gloves or a ruler. But how easily the public was manipulated to voluntarily release that information.

Companies all over the world are ravenous about collecting your personal information. And they are making ongoing efforts to do so. According to the *Wall Street Journal*:

> Many of the most popular applications, or "apps," on the social-networking site Facebook have been transmitting identifying information — in effect, providing access to people's names and, in some cases, their friends' names — to dozens of advertising and Internet tracking companies, a *Wall Street Journal* investigation has found.
>
> The issue affects tens of millions of Facebook app users, including people who set their profiles to Facebook's strictest privacy settings. The practice breaks Facebook's rules, and renews questions about its ability to keep identifiable information about its users' activities secure.... The *Journal* found that all of the 10 most popular apps on Facebook were transmitting users' IDs to outside companies.

This slip spotlights an Achilles heel in the World Wide Web: third-party suppliers and independent software developers. Facebook doesn't produce all its own apps. It buys or leases them from other companies who supply the code, and code consists of volumes of script. A couple of lines, directing the target computer to send postcards to daddy, could easily be missed.

Everyone on Facebook is assigned an ID number, and all your personal information — everything on your Facebook pages — is connected to that number. If someone can get your ID number, they can get your personal information.

The app supplier got the ID numbers of people using the app. With the ID numbers, they could collect other personal information. Then they sold that information to advertising and data firms that track online usage.

After the subsequent brouhaha, did Facebook or other popular websites clean up their act? Hardly. When you press the Like or Tweet button, your identity is matched to your browsing habits on an unprecedented scale and passed on to third-party companies.

This means you're going to have to protect yourself. No one else will, and no one has a greater investment in your privacy.

Protect Your Personal Information Online

Although I do refer to technical issues in this and the next chapter, these sections are still about boundaries, not technicalities. While technical fixes are mentioned, the essence is always about getting clear about what you are trying to achieve. At some point, you'll say, "I can't bother about this," and that is okay.

RULE OF THUMB

- If you convert to Timeline in Facebook, do not provide your birthday, birthplace, or birth year.
- Remember, by indicating your year of graduation from high school or college, you are indirectly revealing your birth year.

Doing something is better than doing nothing. Even if you take just one step, you can still feel good about doing something for yourself.

On Social Media Sites

- Remove personal identifiers from your Facebook page or any other social media site you've joined. Remove your birthday, any information that would reveal your mother's maiden name, and any pictures that would reveal your address.
- Before you sign on to a social media site, see if by default your email address, location information, and name will be shown to the public.
- Check your profile on social media sites that you've already joined. Log out of your account and revisit as a guest or as another user. See what is revealed about you to the public.
- Use privacy settings to screen information that you don't want made available to the public.

HOW TO SCOPE OUT A SOCIAL MEDIA SITE

- Go to the library and use their computer before you sign up.
- Scan the pages of other users of that site as a guest (or, if you are forced to sign up, use a pseudonym and the dry cleaner's fax number until you get the lay of the land).
- Are you seeing the same personal identifiers on each person's page, or do some reveal more and others less? (If other users have a continuum of choices, then you'll have choices, too.)

On Any Site

- When you visit a website, don't automatically sign up; wait until you've decided whether you really want what that website can offer.

RULE OF THUMB

- Your bank can be trusted with your PII.
- Mo's Mudguards may have a lower level of reliability.
- It may be safer to use PayPal or buy through Amazon than give some sites your credit card info.

HOW TO CHANGE PRIVACY SETTINGS ON FACEBOOK

- On the Facebook menu bar, click the arrow on the far right, next to your name.
  - Select *Privacy Settings.*
- You'll see a list of categories for which you can choose settings.
- Click on *Edit Profile* (in first paragraph).
  - Choose *Don't show my birthday in my profile.*
  - Click *Save Changes.*
- Click the arrow at the top on the far right again.
- Again, select *Privacy Settings.*
- Next to Profile and Tagging, select *Edit Settings.* (Tagging means that you or someone else can attach your name to a photo. For a fuller explanation, choose *Facebook Help.*)
  - Decide what limits you want to put around each choice.
  - If you want to review tags before they are posted on your page, turn that feature on.
- Select *Done.*

- Fill out only the required information.
- Think about what information you are giving them.
- How reliable do they seem? (It is not hard to buy a security certificate, so don't assume too much when you see a "trust badge.")
- If a website asks for bank or credit card information, consider paying through PayPal.

Protect Your Personal Information from Certain People

- Do not store your PII or other personal files or email on your work computer. Remember that your boss and coworkers can access everything you do on your work computer (even if they appear to be techno-illiterates).
- If a member of your household is an active addict, do not store financial access information on your computer (or anything else that could be salable).

- If you or a family member, particularly a child, has made a mistake and put too much personal information on your own website, you can take steps to remove that information.
    - In Google, go to http://googlewebmastercentral.blogspot.com/2011/05/easier-url-removals-for-site-owners.html to block a web page from being shown in search results. This lasts ninety days, so you will have to renew it. Google also explains how to permanently prevent a page from appearing in search results.
    - Reputation.com is an online reputation manager that searches the web to find sites that name you and reveal your private information. Once that exposed information is located, you can remove it.
    - Numerous other companies also offer various levels of protection or correction for various fees.
- Notice that you can decide who will see each post on your Facebook wall.
    - Choose your News Feed or, once you've accessed your own Facebook page, click on the word *Facebook* in the top left corner.
    - In the box where you can type your comments, type "test," replacing "What's on your mind?"
    - In the bottom right corner, next to *Post*, there is a grayed icon. If you see a globe, it means your post will be visible to the public.
    - Click the down arrow to designate whether your post can be viewed by the entire world, friends, or just close friends.

HOW TO CHANGE PRIVACY SETTINGS ON LINKEDIN

- Choose *Settings*.
- Select *Profile*.
- A box opens headed *Privacy Controls*.
- Select each option, choose your preference, and hit *Save*.

- You can go through every listing on your site and designate a particular privacy setting.

Sharing Your Computer

If you must share a computer with someone else, it's important to consider every aspect of privacy, exposure, and parameters around browsing and types of websites visited. You can set boundaries around what the other person can and can't do. Think about and then clearly state what the limits are. If necessary, discuss or negotiate these with the other users and come to a clear agreement.

(Some computers make it possible to create different accounts, so that each user has their own set of files that are not accessible by other users. If your computer has this feature, there's no problem; you can skip to the next chapter.)

Here's a checklist to discuss or fill in:

Boundary Checklist for Other Users of Your Computer

- ❒ Browsing permitted.
- ❒ Browsing limited to:
  - ❑ YouTube
  - ❑ Twitter
  - ❑ ______________
- ❒ Browsing forbidden to:
  - ❑ MySpace
  - ❑ WeirdGames
  - ❑ ______________
- ❒ Computer camera stays off at all times.
- ❒ Computer camera limits:
  - ❑ Face only
  - ❑ No body parts below neck
- ❒ Downloading permitted.

- ❒ Downloads limited to:
  - ❑ Receipts
  - ❑ Articles
  - ❑ Instructions
  - ❑ ______________
- ❒ Downloads forbidden:
  - ❑ Bundled music
  - ❑ Freeware
  - ❑ Apps without my permission
- ❒ Uploads forbidden:
  - ❑ Any photos showing our family or home
  - ❑ Any photos or videos revealing our address or location
  - ❑ Any of my personal emails
  - ❑ Any of my work product
  - ❑ Any of my financial information
- ❒ Uploads limited to:
  - ❑ Nature photos
  - ❑ Car photos
  - ❑ ______________
- ❒ Installation of programs, applications, or apps not allowed.
- ❒ Installation of programs, applications, or apps require my permission first.

Of course, add any other limits you want to set around uploading, sharing your information, changing your preference settings, adding things to your desktop, or using a coaster under a dripping glass on your black cherry desk.

Let other users know the consequences of a boundary violation in advance. It is always easier to follow through with a consequence if the other person knows what will happen and makes a choice despite that knowledge.

For example: "As I told you, you had my permission to use my computer, but only as long as you respected the boundaries I set for its use. You chose to download porn, so you will no longer be allowed to use my computer. I will now put password protection on my computer. You'll have to go to the library."

If you use someone else's computer, ask them to fill out the same checklist so that you'll be very clear as to what you may and may not do. Then, of course, respect those boundaries.

If you jointly own a computer and you each have a different degree of concern for your online privacy, use separate browsers. Use separate email managers, too. Then you can each set preferences to your level of comfort. If your computer allows you to color your files and folders, you can each have a particular color that you use for your own material.

If you have no choice but to share your computer with someone you can't trust to honor your boundaries (or stay out of your private files), keep all your sensitive files (and all passwords) on a flash drive in your pocket.

PART IV

CYBERSPACE TRICKERY

Chapter 13

SPIES

Microbeings crawl the Internet, some getting into your computer. You can set boundaries with some of them. You can't do anything about others.

If you already feel up to your earlobes with techno-talk, skip this chapter for now and come back when you're ready to advance to another level of self-protection.

If you do decide to read on, and you reach a point where your brain fizzles, skip forward to the following chapter. We all reach a level of saturation with techno-speak, especially those of us who (like me) are non-geeks. Don't force yourself. Let your neurotransmitters replenish themselves. Respect your own endurance boundaries.

Creepy Crawlies

I remember thinking naively that there's so much data racing through the sky that my own unimportant missives would get lost. But sophisticated software can lock onto certain key words in emails and blogs. Crawlers, spiders, and bots can creep through websites, pick out the items that relate to their agendas, and pass them on to the people or companies that can use them.

And then there are the little tattletales that shadow you while you, in the privacy of your home or office, are under the illusion that as you go from site to site, you are alone.

You are not alone. You are being tracked.

Here's what they do, how they do it, and what you can do about it.

Cookies

A cookie is a packet of data that is sent to your computer by a website and then returns information back to that website. It can do a variety of things that make your connection with the site more efficient and personal. It can authenticate your right to sign in, communicate your preferences, and identify your computer. It might also track you as you browse the Internet.

Let's say you go to happywidget.com and put two red widgets and one yellow widget into your shopping cart. Then your sister calls to tell you she's bringing you fresh summer squash right out of her garden. You are so crazy about squash that you turn off your computer and forget all about your cart of widgets. No problem; next time you go to happywidget.com, your cart will be patiently waiting, still holding your colorful widgets.

How does it remember? A cookie. Your computer doesn't actually hold a microscopic shopping cart with miniscule widgets. Instead, the cookie grasps a unique session identifier (an ID number) that links to your name, your user name, and your billing address.

That's how the store recognizes you when you return and happily says, "Hi, Jane! Welcome back!"

This seems harmless enough, but now imagine: In your own hometown, you step into a sock store and pick up a pair of purple argyle socks. You decide not to buy them and leave the store without a purchase, then walk down the street to the drugstore. Just inside, you find an entire display of purple argyle socks. Spooked, you leave and pop into the coffee shop across the street. The barista offers you today's special: a cappuccino plus a pair of purple argyle socks for $9.99.

Incredible coincidence? Or did the sock store manager race across the rooftops, informing retailers willy-nilly about your interest in purple argyles? In the brick-and-mortar world, this would never happen; in cyberspace, it happens constantly.

Why don't we question online situations that would have us bug-eyed on Main Street? Because the Internet is so boundless. It's hard to find a frame of reference in an infinite universe. It's difficult to get a feel for what the normal boundaries would be.

When the argyle equivalent happens online, it's no accident. That's what cookies do. You are being shown a targeted ad. And the further you drift through the Internet, the more precisely the ads will reflect your interests. How does this happen?

While you're visiting the widget store, a banner ad for All Things Red, one of happywidget.com's advertisers, appears. At the same time, All Things Red also sends your computer a cookie. This is a third-party cookie. This cookie will sit in your computer, and when you visit a second website, the All Things Red cookie will throw up an ad that relates to this new site or that reflects your interest in red things or both. And your computer will pick up more cookies.

As you click through the Internet, a trail of cookie crumbs follows you. (Are you starting to feel like Hansel or Gretel?) The further you go, the more the ads will relate to your interests, based on the previous sites you've visited.

Viewing a single site can generate hundreds of tracking files, sending information about you to hundreds of companies. Cookies can be intercepted in transit while you pass through a website that is not secure. A hacker can steal your computer's cookies and gain access to your accounts.

To find good information about cookies, take a look at the *Wall Street Journal* series entitled "What They Know," at http://online.wsj.com/public/page/what-they-know-digital-privacy.html. (As mentioned before, if this URL brings up the message "page not found" or something similar, go to wsj.com and enter *cookies, what they know*, or *digital privacy* into the site search box.)

Web Bugs or Web Beacons

Web bugs or beacons are invisible electronic tags that can talk to any of your cookies that were issued by the same company that owns the bugs. They can access your personal information if you register for an online service or enter a contest.

Suppose you have a toe rash and visit toerash.com, looking for help. As soon as you arrive at the site, a new cookie is planted in your computer. Then, a few sites later, at happynailbiters.com, a period-sized cookie-detecting web bug will say, "Hi, Cookie, who made you?"

The cookie replies, "Tinyclicks.com."

"Me, too! Where've you been?"

"Lots of places. Let me show you all the websites I've visited. I've gone to toenailrashremedies.org, toeprobs.us, toetalk.com, exoticrashes.biz, highkickingmamas.fun, weirdtoewear.org, and sexytoes.you."

"Cool. I'm just gonna pass that on to our creator at Tinyclicks."

So, if you were hoping no one would find out about your toe rash, it's still a secret — just between you and everyone who knows how to access the register at Tinyclicks. Did you sign up for the free

ToeNail News Tips and leave your email address? Then that cookie number could be linked to your personal identifiers.

What happens if Tinyclicks is bought by GiantSearchEngineIn-TheSky? I'm sorry to say, lots of people now know about your toe rash. Better come clean about it on your eHarmony.com page.

Have you ever gotten a bulk email letter, one sent to thousands of people? You've probably been bugged. Your email address could be encrypted so a bug could tell the company that you opened it, that you read it three times, and that you forwarded it to your cousin.

Can you see this bug? No. Can you detect it? No. Does the website owner have to tell you it's bugging you? No.

If a cookie-tracking law is passed, will bugs be mentioned? We'll see. It's another reason to tune in periodically to the boundaries world.com website.

Are you responsibly blocking cookies and washing your cache? (Don't worry if you don't know how. That's just ahead.) If so, that's good, but it won't make any difference to these waterproof bugs. They can still collect your IP address, ID your browser and operating system, and track your movements across any site that carries related bugs.

A beacon can log any text you type on a website. What do you type? Your password, your credit card numbers, your address.

This means that you have to trust a big business to not be greedy, stay ethical, be responsible, and be vigilant about checking the work of all its programmers and third-party suppliers. Feel better?

Spyware

Spyware transmits information about you to someplace on the Internet, without your knowledge. Spyware tracks your website visits and uses that information to build a profile of you, but some can also collect your personal information and even change the configuration of your computer.

Where does spyware come from? It can be bundled into free software or into music or video file-sharing programs, with embedded instructions to collect your personal information. You might be notified if you read that really long Terms of Service agreement that you have to click before you get something for free. Otherwise, if you — like most of us — click without reading every annoying word, you will experience the truism that there is no free virtual lunch.

Adware

Some tablet computers and e-book readers are available in cheaper versions that come loaded with adware. If you're willing to put up with constant display ads, you can save considerably on the purchase price.

Adware can also be installed as a part of software you download. Then it installs commercial links and ads onto websites you visit, sponsors pop-up ads, and could install spyware. The money saved by bundled software or freeware may actually cost you, if malware is sneaking into your machine at the same time.

What They Know

The *Wall Street Journal* analyzed the tracking files installed on people's computers by the fifty most popular US websites. It then built an "exposure index," which measured how much each site exposed visitors to monitoring by cookies, beacons, and other tracking technologies. Here are some of the findings about website viewers' exposure to tracking:

Very High Exposure

- dictionary.reference.com

High Exposure

- msn.com
- comcast.net
- merriam-webster.com

Medium Exposure

- ask.com
- yahoo.com
- answers.com
- walmart.com
- imdb.com
- att.com
- amazon.com
- eBay.com
- chase.com
- yp.com
- aol.com
- verizonwireless.com
- cnn.com
- go.com
- myspace.com

Low Exposure

- bing.com
- weather.com
- mapquest.com
- youtube.com
- mozilla.org
- twitter.com
- apple.com

- paypal.com
- target.com
- linkedin.com
- bankofamerica.com
- microsoft.com
- adobe.com
- Craig's List

No Exposure

- wikipedia.org

The *Wall Street Journal* study found that dictionary.com installed 234 trackers; AOL, 133; MySpace, 108; Bing, 59; Adobe, 30; Craig's List, 4; and Wikipedia, 0.

Who is watching? Facebook, Apple, Disney, EBay, LinkedIn, Experian, Amazon, Microsoft, Google, and Yahoo, to name a small slice of the companies that use the information collected by the tracker files.

So if, on Saturday morning, you visit ten of the top-visited websites, each of which install an average of 140 trackers while you're there, by noon your computer is holding 1,400 trackers, all of them broadcasting your browsing trail to multiple companies, sometimes through a "middleman" company that compiles the data and forwards reports to subscribers.

Here's an example of what tracking files can do. (Two independent investigations uncovered similar results.)

> The *Journal* found that Microsoft Corp.'s popular Web portal, msn.com, planted a tracking file packed with data: It had a prediction of a surfer's age, zip code and gender, plus a code containing estimates of income, marital status, presence of children and home ownership, according to the tracking company that created the file, Targus Information Corp.

> Both Targus and Microsoft said they didn't know how the file got onto msn.com, and added that the tool didn't contain "personally identifiable" information.

But we've already learned from our smart Harvard scientist, Dr. Latanya Sweeney, that with a person's age, zip code, and gender, one is well on the way to identifying a specific person.

Do you feel better knowing that Microsoft and Targus didn't know how that tracking file got onto msn.com? I don't. Either they don't know what their programmers are doing, they can't monitor third-party suppliers, or they aren't being completely honest. Regardless, we aren't protected.

Are we safer now, two years later?

An article in the *Wall Street Journal* describes the experience of a man in Georgia:

> He sent a note to a showroom near Atlanta, using a form on the dealer's website to provide his name and contact information. His note went to that dealership — but it also went, without his knowledge, to a company that tracks car shoppers online. In a flash, an analysis of the auto websites [he] had anonymously visited could be paired with his real name and studied by his local car dealer.

When this man next walks into a dealership, the salesman will know how to approach him. The article goes on to say:

> The use of real identities across the Web is going mainstream at a rapid clip. A *Wall Street Journal* examination of nearly one thousand top websites found that 75 percent now include code from social networks, such as Facebook's "Like" or Twitter's "Tweet" buttons. Such code can match

> people's identities with their Web-browsing activities on an unprecedented scale, and can even track a user's arrival on a page [even] if the ["Like"] button is never clicked.
>
> In separate research, the *Journal* examined what happens when people logged in to roughly 70 popular websites that request a login and found that more than a quarter of the time, the sites passed along a user's real name, email address, or other personal details, such as username, to third-party companies. One major dating site passed along a person's self-reported sexual orientation and drug-use habits to advertising companies.

So despite growing concern, probing studies, and savvier Internet users, this horse is running away from us.

Other Frauds and Dangers

We all know what spam is — a favorite Hawaiian meat. It's also UBE — unsolicited bulk email. Current estimates are that 88 to 90 percent of emails flying through the ether are UBEs. It's annoying. It takes up space and the time required to delete them. Are there other dangers?

Yes, bulk email can be filtered through zombie computers that contain worms that allow the spammer access to your computer. A worm holding malicious code can enter a computer and direct it to go to a website and download a Trojan horse. Then the planted program can disable the computer's firewall and its antivirus software and open a back door, through which it can be controlled by a remote computer. In other words, a remote person can make your computer do bad things.

Even though this sounds like a *Twilight Zone* episode, your computer could have a secret life you know nothing about. Some thirteen-year-old in Brazil could be playing your computer like a piano, although most zombie computers seem to actually be in companies and universities that have no knowledge that their machines have been taken over.

Big companies have spent big money getting you to trust the Internet, but the reality is that this is a house of cards held together with naïveté and trust. We hold our breath, take our chances, and hope for the best. We are all so dependent on computers and the Internet, we can't stop just because it's dangerous. We're like the brave souls who boarded the *Mayflower*. They didn't know where they were going; they didn't know if they would survive when they got there; but once they were halfway across the Atlantic, there was no turning back.

And who is the new target? Sneaking trackers onto smartphones by secretly bypassing privacy settings. (Google has already done this.) Pretty soon that brain-dead land line and the old Smith-Corona may seem downright cozy.

What's a Simple Non-Geek to Do?

Solutions are available, but some of them will lessen your online convenience. You'll have to decide where you want the balance to be — in the direction of greater privacy and more vigilance and inconvenience, or more convenience and less privacy. Below are some simple and effective ways to protect your computer and set boundaries with electronic intruders.

Manage Cookies

Begin by using the online Help function within your browser (Firefox, Internet Explorer, Safari, etc.), or look under the Privacy and/or Security headings.

- *Delete cookies from your browser:* In the Tools menu, look for such phrases as *Clear Recent History*, *Delete Browsing History*, or *Show Cookies*.

- *Use private browsing:* Look under Tools for phrases like *Private Browsing*. This allows cookies to be installed while you are on the web but automatically removes them when you close your browser. (The browser can't discriminate between good and bad cookies, so all from that session are ditched.)
- *Block the installation of cookies:* Look under Tools or Preferences for such phrases as *Accept Cookies* or *Block Cookies*. Each browser also gives you the option to block the installation of cookies from specific sites (see your browser's online Help for details).

Install Updates

Legitimate companies hire good people to untangle the newest threats and figure out how to disable them. These then become part of the free updates to existing programs that you are offered periodically. Accept these updates; they usually take only a minute or two to download. (These are different from *upgrades*, which are new versions of programs; these often cost money and usually take much longer to download.)

If possible, turn on Automatic Updating.

Don't Open Email Links

So far, email itself can't harm your computer. However, clicking on links within email can cause problems.

If you receive an unexpected attachment from someone you know, and anything about their email seems odd (for instance, if the email says something very general, like, "Check this out, bro, you'll love it"), don't open it. First check with the sender to make sure they sent it. Their email account may have been hijacked.

Similarly, if anything about an email seems odd or generic, and that email contains a hyperlink (a link to a web page), don't click on that link. Check with the sender to make sure they sent the email; email addresses can be stolen or hacked. A hyperlink can open a spyware scam or take you to a web page that automatically downloads a virus.

Only install add-ons that you trust. Web companies can offer free doodads for your computer that will install spyware or other malicious content.

Protect Your Finances

When you are finished visiting your online bank, credit union, credit card company, or PayPal, be sure to log out. Then close that browser page. Then close your browser.

If you've set your privacy or security settings to delete cookies when you close your browser, all those implants will be erased when you go back online, and cookies related to money and banking will be gone.

Advanced Moves

- Learn who is tracking you and then block them with ghostery.com. This free tool gives you a roll call of every company that is tracking you; enables you to learn more about each one; and gives you options for blocking or deleting items.
- Block many online ads by using the free tools at privacy choice.org or networkadvertising.org. These tools use cookies, however, so you will have to reinstall them whenever you delete cookies.
- Check your computer's operating system to see if an unwanted program is running in the background. In Windows, do this through Task Manager; in Apple operating systems, do it through Activity Monitor.

- Detect adware and spyware
    - Be careful what you download.
    - Install a firewall that asks your permission before it sends or receives data.
    - Use an antivirus program that has automatic update capability.
    - Use an anti-spyware program especially designed to detect, quarantine, and remove spyware.

The Ongoing Cost of Internet Efficiency

The above remedies are a way to create electronic boundaries to protect your privacy and safety. Some of these won't cost you much in time. If you set aside an hour now to get comfortable with the available safety features, ongoing maintenance in the form of updates and cookie-cache emptying will take a small amount of regular additional time — perhaps fifteen minutes to half an hour per month. You will have the satisfaction of knowing that you have taken reasonable precautions.

Still, the very fact of high web use — requiring ongoing maintenance in the form of monitoring and cookie sweeping — can lead to a violation of your time boundaries if this time is being taken from your productive hours, your family time, or your time for yourself.

The first year I did all my Christmas shopping online, I remember remarking, "I saved so much time." That was then and this is now. For some people, the Internet may not be such a time saver. Peacefully taking a bus to the store, meandering among the aisles, making your purchases, and reading while the bus takes you home may end up being a far more efficient use of an afternoon.

If you decide to sign up for web-monitoring services, pay attention to the full cost of using the web in terms of time, energy, and money. When faced with a task that you can complete online or in the flesh-and-blood world, compare the likely costs of each. The web may or may not be your best option.

Chapter 14

DON'T SET YOURSELF UP FOR TROUBLE

This chapter pulls together warnings from individual chapters to show how some online situations can combine to cause you trouble. For example, here I am, minding my own business, writing this book on how to set boundaries in an overconnected world. I go to a social media site to see if my privacy settings are up-to-date. An ad in the right column says, "Save 50 percent on auto insurance if you only drive within 30 miles of your home. The regulation that insurance companies don't want you to know."

And I fall for it, despite my advice in this very book about holding a boundary against commercial interruption while working. I click on the ad. It promises to give me the relevant tip — after I fill in my email address and name. I pause and think about how good an email filter I have — that I only have to label any incoming email as junk, and my computer will remember it. So I fill in my info.

The tip is not revealed. I just have to fill out one more page with other little bits of personally identifiable information. I fill out the page, using a fax number (rather than giving them my phone number) and a post office box address (rather than my home address).

The tip is still not revealed. The next page asks for annual household income and the ages of my household's occupants. I give them a pitiful annual income to demonstrate how useless it is to come after me, and the age of my dog.

The whole time I'm doing this, I'm fully aware that I should not be doing this, that I am falling for a pitch, and that I should not be revealing any of this information. But I want that tip.

I finally finish the fourth page of the form, where I list my interests as reptiles and types of moss, and — waiting breathlessly — discover that the tip is not forthcoming.

Within less than a minute, four insurance companies send me emails. I send an outraged reply to each one, describing the ad, whining that I didn't even get that tip, and telling them to never contact me again. From one, I get an apology and a promise to not contact me again. From the others I receive nothing.

I still haven't gotten that tip.

I doubt that I'll fall for a pitch any time soon, but this incident does demonstrate how even an expert can be tricked if an ad promises something so good that people slide their doors open.

Five Ways We Unwittingly Weaken Our Boundaries

Pitfall 1. Falling For an Ad That Says Just What You Want to Hear

Filling out a form that reveals your email address and other personally identifiable information is like opening a gate in a fence; it lets the wolves into the yard. It also creates an exception for this particular brand and for the parent company that owns the brand, even if you've registered with the www.donotcall.gov list. Now this company can legally call you whenever it wants. (And if you give them your cell phone number, they can call or text it, too.)

Commercial companies know what you like — based on the information they've gotten from trackers that follow your progress through cyberspace — so the ads that appear on the pages you surf will be slanted toward your interests. (Somehow, they know I like a bargain.)

All they have to do is put together your mania for birds, your fascination with do-it-yourself projects, your love of cute miniatures, and your favorite color (yellow) to offer you a DIY yellow birdhouse with little window boxes and micro-black-eyed Susans. And you're theirs.

You'll even get a second one free. All you have to do is pay shipping and handling — on both of them. Can you opt out of the second birdhouse? No. You rationalize that it will make a good present for Aunt Feathers. You are then offered a year's supply of birdseed, a squirrel-proof seed-feeder, and a retractable bird-feeding pole. You turn them down.

But wait, you can also get a Sexy Beaks fifteen-month calendar and an alarm clock that chirps you awake.

That little twenty-second interruption has now eaten twenty minutes of your concentration. By now you've invested too much time to shut the thing down. You keep thinking you're almost done, when another page offers you some other splendid thing.

Ideally, the product will live up to its promise. But if it doesn't — or if, like me, you get sucked into filling out a form and then the promise isn't delivered on — protest. Make your objections to any business that pays into or sponsors that ad.

Also, be savvy about what you put on a form. Don't just automatically fill in every blank. (I wonder how soon I'll be seeing ads for moss and reptiles.)

Pitfall 2. Falling for Official-Looking Email Scams

I recently received a suspicious-looking email, purportedly from the US Postal Service. It said: "Unfortunately we failed to deliver the postal package you have sent on the 27th of June in time. Because the recipient's address is erroneous. Please go to the nearest UPS office and show your shipping label," and included a bar code and a URL link to click on. Two things alerted me that the "usps.com" message in my inbox might not be a legitimate email: 1) poor grammar and 2) knowing that no package I had sent from my post office had my email address. So I took a screen shot, printed it, and took it to the post office. They scanned the bar code and verified that this had nothing to do with the birthday gift I'd sent to my mother.

When I got back from the post office, in the interest of doing research for this book, I typed the web address into my browser letter by letter (instead of clicking on the URL link in the email header, since a spybot could be tucked into the box labeled *Print a Shipping Label*, or a link could be embedded in the entire label). I was directed to a page with a warning from the postal service that the address might be a scam and asking if I actually wanted USPS.

This is a well-put-together scam, and further research informed me that clicking the label does install malware that will attempt to access data on your computer. The senders probably know that I opened the email, and now my email address is probably on some list. I'll notice if I start to get other unknown emails with links in them. However, by not actually clicking on any included link, and by taking a screenshot, I have probably averted any serious intrusion.

Pay attention to poor grammar, tenses not matching, spelling mistakes, or something not quite making sense — like knowing the agency in question would not have your email address. My label shows USPS icons, but in the instructions I'm told to go to UPS to follow up. I doubt UPS handles US Postal Service delivery issues.

Another round of current scams gives the appearance of being sent by the IRS. The wording is such that, in your sudden anxiety at being on the wrong side of the taxman, you might hastily click the link to see what kind of trouble you are in. You can go to the www.irs.gov website and find out what current IRS-related scams are traveling the net. But, remember, if the IRS wants your attention, they will send you a physical letter.

I believe that we are all more vulnerable to scams because we have been trained by our computers to do things quickly. Plus, having so much clutter in our email boxes forces us to sift through them as rapidly as possible. We want to quickly dispense with unasked-for intrusions. In the midst of this hasty handling, we are more likely to open an official-looking email without even thinking about it.

Don't automatically click inside it. Stop and think about it first.

(This is another cost of electronic clutter — being required each day to handle multiple intrusions, to discriminate among them quickly and superficially, and to discard them — all to prevent a Trojan missive from grabbing our privacy.)

If at all possible, have one computer for email and browsing, and a different, non-networked computer for all your private information and work.

Pitfall 3. Thinking That Cyberspace Is So Big, You Won't Be Noticed

The world has so many roses, an aphid won't notice mine. But there are plenty of aphids to go around, just as there are plenty of web crawlers and other tiny bits of software that mindlessly follow their master's instructions, scanning millions of pages of code in the blink of an eye.

Thousands of computers are interested in learning more about you, so think about what you share, post, send, attach, and accept. If you tend to reveal all, occasionally tune into chapter 12 or go to www.boundariesworld.com.

Pitfall 4. Thinking That Cyberspace Is So Big, You Are Helpless against It

Both the fascination and the peril of the Internet are that its workings are invisible to ordinary people. It is so removed from flesh-and-blood, bricks-and-stone life that we have to work to embrace a realistic perspective on it.

I was in a computer store, learning about the innovations in their latest laptop. The salesman was talking about the cloud — how information could be stored there and magically appear on all my digital devices instantly, whenever I changed a tiny thing, like a calendar date, on just one device.

I said, "Where is the cloud?"

He said, "On the Internet."

"Yes," I said, "but the Internet is composed of many, many computers all linked to one another. Where is the cloud actually located?"

He looked puzzled. He looked up as if all the data being transferred among millions of devices were simply floating up in the air somewhere. He said, "In cyberspace."

"But there is a building somewhere, that is holding a machine that is collecting and sending out this information," I said.

He looked so lost, I almost felt sorry for him. I added kindly, "Cyberspace is a metaphor, and so is the cloud. These are metaphors that represent the computers, cables, servers, and linkages that make it all happen."

He said, "I'll go ask somebody."

The salesman was young and had never known a world without nearly magical computers, an invisible Internet, and instant communication. He'd accepted the metaphor as the reality.

We know that the pictures and words we see on our computer screen are courtesy of pixels and code; we know that an engine in the background is making it all work; but, unlike with our car's worn-out air filter, very few of us know how to open the hood and replace the filter on a server that is sitting in India. And most of us can't cybernate (hibernate from cyberspace). Those of us who've gotten used to getting things done on the Internet don't want to go back to snail mail, heavy reference books, or lengthy days in a research library.

Cyberspace *is* big. But that doesn't make you powerless. Real companies in steel-and-concrete buildings, with real people, are behind everything that happens.

Follow the advice in the previous chapters about protecting yourself and your computer. Keep in mind that someone wants your dollar. Take reasonable precautions. And, if you really are hurt by something someone does in cyberspace, go to the company that brokered the deal. Find the breathing person at the place that sponsored the ad or the product. Go on the Federal Trade Commission website (www.ftc.gov) and search for *consumer protection*, *consumer information*, or *scam alerts*. Also check the *Wall Street Journal* (www.wsj.com) or consumer magazines or newspapers.

It is important that we stand up and fight for ourselves, so that companies are forced to remember that we are real people, not just wallets for them to pick.

Pitfall 5: Not Being Aware That Your Need to Be Seen in a Certain Way Can Create a Breach of Your Boundaries

I lead a group in cyberspace. The women in this long-term group have top-notch skills, having learned processes that give them peace with their feelings and ease in exploring the deeper (sometimes painful) parts of themselves. They work together brilliantly.

One day, a group member (I'll call her Agate) asked, "How would you all feel about changing the group? I was talking to a friend and, knowing how beneficial this group is and feeling compassion for her, I wondered about inviting her to join."

The group kicked around the proposal and looked at various

aspects of the issue, using a set of skills that allow them to explore differences without getting polarized. Ruby was completely supportive of the idea. Then she described a similar discussion in her cooking club. There, she had been in full support of an idea — until others raised issues that she hadn't thought of herself, but that she could see once they were stated.

It was an interesting confluence of comments from Ruby. Perhaps she was doing the same thing again here: voicing total support while not exploring her own reservations.

When no one else brought it up, I asked, "Agate, I noticed that when you proposed expanding the group, your focus was on helping your friend. I'm wondering if the primary impetus for you is taking care of your friend?"

Immediately Ruby said, "I wondered that, too."

Agate explored her own impulse to caretake, and then I asked Ruby, "What was going on, that you didn't mention your concern that Agate was caretaking?"

Ruby said, "I wanted Agate to do it for herself. I was worried that if I brought it up, it would seem like I had a hesitation. My impulse was to show that I was really supportive."

This scenario has wisdom for other virtual gatherings, especially private chat room groups and online support groups. We can have an engine inside ourselves that will invite others to break and enter, whether we're standing with someone in the school parking lot, blogging, posting on a Facebook wall, or sitting with virtual friends in a virtual coffee shop.

We can all, often without knowing it, expose what we're trying very hard to prove to the world. We want strongly to be seen in certain ways: as truthful, not cagey; as generous, not Scrooge-like; as helpful, not selfish; as independent, not needy; or as smart, strong, rich, or successful, rather than average, weak, struggling, or a runner-up. And we may be on such a mission to prove these things about ourselves that we miss looking at what we *ourselves* need out of a situation.

In cyberspace, we can be so focused on our own mission that we don't see what else we are broadcasting. It can also prevent us from keeping careful boundaries in online social situations.

Plus, acting out of that drive can land us or our online hangout in a sticky situation with an unsuitable intruder, causing either the death of the group or, at the least, a large interruption to the usual business of the group as we are forced to handle the sticky situation.

As Ruby explored further, she realized she had not identified what *she* wanted, and she gave herself room in our group to explore how her mission can lead her to slight her own needs. She soon recognized a similar pattern with her husband, being so concerned that he see her as helpful, that she had ended up with more responsibilities than any one person could carry.

It can be easier to figure this out by looking at your off-line conduct first. Therefore, this exploration will proceed in two parts: off-line, then online.

Part 1: Explore Your Off-Line Pattern

Do you try to prove something about yourself in the real world? What is it you want people to see in you? What are you trying hard to not be? You might get some clues about this if you have a parent who does (or did) things you don't admire.

For example, my mother was always concerned only with herself, so for years I worked to prove that I'm not like her. As a result, I've sometimes given away far too much of myself. Once I understood this, I could begin to catch myself whenever I fell into that old pattern.

After you discover your own desire to be seen in a certain way, and an automatic behavior pattern that springs from it, you can then create boundaries that protect the person you want to become. Those boundaries will be necessary, because people may not applaud your change if it affects the way you interact with them.

This boundary-setting process includes six steps.

1. Start noticing your automatic default setting in your relationships. Do you tend to automatically:
   a. Be too open or closed?
   b. Give too much or too little?
   c. Try to make others feel a certain way?
   d. Come on too strong? Hold back too much?
   e. Try to be noticed for something?
   f. Try to be seen in a certain way?
   g. Pretend to be something you're not? If so, what? To whom?
   h. Try to create a certain type of bond, or relationship, or ambiance?
   i. Try to re-create a previous situation or time in your life?
   j. Try to change the world? How? (By the way, there's nothing wrong with trying to change the world. These questions are about your default settings, not things you do wrong.)
2. Think about what you may be missing because of these default settings. Do you:
   a. Get taken advantage of?
   b. Take too much from others?
   c. Keep others away, or lose them entirely?
   d. Let the wrong people get too close?
   e. Get enmeshed in other people's lives in unhealthy ways?
   f. Not take care of your own needs?
   g. Limit your range of activities and experiences?
   h. Not get enough time alone?

   i. End up diluting the depth of connection that could be possible for you?
   j. Not pursue your own deepest interests?
3. What new paths do you want to create for yourself? What new behaviors and activities do you want to try? What aspect of yourself or your life do you want to expand?
4. Set (or reset) a boundary that gives you something you've been missing or protects a new path. (See chapters 3 and 8 for details on this process.)
5. Expect to be accused of being exactly the type of person you've always tried to show you aren't. (For example, if you've spent a lifetime trying to prove you are unselfish, the first time you stand up for something you want, you will probably get called selfish and thoughtless — possibly by someone you love. This is an inevitable part of the process.)
6. Don't weaken your boundary in response to #5. Instead, strengthen it. At the same time, value the accusation because it gives you the opportunity to:
   a. See that you are really making a change.
   b. Learn something about the other person, and about the type of relationship your default settings have made you vulnerable to.
   c. Perceive how the other person may have been using or manipulating you, or making the most of your default settings.

For some people, their use of you has been unconscious. They've simply enjoyed the fruits of your mission without questioning it. Others may well have seen that this was your Achilles heel and used it deliberately. In either case, you'll learn a lot about your friends, and you'll learn a lot about your relationships.

Now, let's take this awareness to the Internet.

PART 2: PROTECT YOUR ONLINE HAUNTS

If you are in a private support group, or meet with a passel of friends in cyberspace, when do you open the group to others, and when do you protect its boundaries?

You can all learn a lot by doing the part 1 exploration together. In fact, if you have already picked up on a fellow member's default settings, you may be able to offer them insights that they haven't been able to discover for themselves. Other group members will probably be able to do the same for you.

As you learn more about your tendencies, both individually and as a group, you'll be able to sense when a suggestion to open the group's boundaries comes out of some of these automatic responses rather than a genuine desire to strengthen or improve the group.

This doesn't mean that the group should never open its boundaries to others, but before doing so, the members should decide what questions it wants answered about a potential new member. Then you can make an informed group decision together.

If you do decide to open the door to others, set up a trial period for any new person, *in a different chat room than your usual meeting place*. That way, if the person turns out to be a poor match, they won't know how to access your usual meeting place.

What if you've already let someone in and they aren't working out? Be direct and tell them so. Ask them to stop attending your group. Depending on how the online chat room is set up, you might have the capacity to block unwanted visitors.

However, if there's any possibility that someone could enter your chat room without being detected, you will have to set up a new room with a new password. You'll only be able to do this if you already have each member's private email address, so you can email the new chat room location and password to the members you want to keep.

Groups with Leaders

A group with a trained leader has an additional margin of safety, as protecting group boundaries is part of a leader's job.

In my own online group, I serve as the leader so I have the most responsibility to protect the integrity of the group. When Agate talked about her friend, she used an occasional word that told me there might be a problem with the potential new member. I created a little test that allowed me to see if my hunch was correct. My test worked, and the person did not join. (I can't describe the test, in case I need it again. It has to do with suspecting that the unknown person had a personality disorder; then, in our phone interview, I set an entry task that, if I was correct, a person with that particular personality disorder would not want to do. In that way, she self-selected herself out of the group.)

If you are a group's leader, remember that you have your own default settings and tendencies. Like the members, you will benefit from the exploration in part 1. It can be immensely useful to realize just how much your unconscious leanings may be influencing your group.

As I've said before, boundaries are isomorphic. As they are strengthened or made more flexible by a group's leader, that change morphs through the group as well. In fact, during the forty years when I was a therapist in private practice, I was amazed that every time I freed an issue in my own individual therapy, my clients would all, magically, come into the office with the very same issue within a matter of weeks.

PART V

INTIMACY

Chapter 15

FRIENDS VS. "FRIENDS"

A man must eat a peck of salt with his friend before he knows him.
— Miguel de Cervantes, *Don Quixote*

The art of life is to keep down acquaintances. One's friends one can manage, but one's acquaintances can be the devil.
— Edward Verrall Lucas, *Over Bremerton's*

We used to understand the difference between an acquaintance, a friend, and an intimate. Hopefully, in our embodied lives, many of us still do.

That these distinctions are now blurred in cyberspace indicates how we grope to find perspective in a universe that has no grid-lines. We call it a web, but we can't see the strands.

This lack of benchmarks burgeons on social media sites. A Facebook "Friend" may be a distant acquaintance. We may not even know some of our Friends at all. Yet we may be revealing our politics, our children's names, even our tender moments to a host of strangers.

Lucas's observation that one's acquaintances can be the devil can still apply today, though he, bridging the previous two centuries, had in mind persons he knew by sight. We no longer know our acquaintances by sight. Nor do we know their agendas, their perversions, or their true identities.

Some folks try to build their Facebook Friend lists. They like to boast of having five hundred Friends. But you know that these are not, all of them, true friends, right? A significant percentage are actually strangers.

> RULE OF THUMB
>
> **Separate Acquaintances from Actual Friends**
>
> 1. Click on your name on the top banner in Facebook.
> 2. Click *Friends* in the list under your photo.
> 3. For each person on your Friend list, choose whether that person is a close friend, friend, or acquaintance.
> 4. Since you aren't given an option to say someone is a stranger, designate as acquaintances those people you do not actually know.
>
> Now when you post something on your wall or timeline, you can choose which category of friend can view it.

It's important, as you accept Friend requests, to keep your perspective on the difference between someone you know and someone who is a stranger. (Facebook does allow you to classify the people on your list, and I encourage you to do this.)

Designate as close friends only those you truly know and trust. Strangers should be marked as acquaintances. If you plan to reveal all your thoughts to Friends (in addition to Close Friends), remember that by selecting *Share*, they can put your comments on their walls, where anyone on their list can see what you said.

(Test it yourself. Go to a Friend's wall or timeline, and choose *Share* under something on their list. Voilà, it appears on your wall.)

Do you know how discriminating your Facebook Friends are about their Friends? Do you know if your Friends actually know and can vouch for everyone on their own Friend lists? Do you want to take the time to research this question with every person on your list?

Anyone who comes to your Facebook page can click on *Friends* and see all the names (and pictures) on your Friend list. That means that anyone who visits your Friends' lists can find you (and your picture) there.

It's a bit mind-boggling, like looking into a mirror that is reflecting another mirror. The mirrors echo themselves to infinity. The possibilities of finding and being found stretch as endlessly.

Keep in mind that not everyone is who they claim to be. Some people may not be real people at all, but mannequins run by an individual or group with commercial (or even less noble) interests.

And remember: You run the same risks on online dating sites.

Chapter 16

ONLINE DATING

Looking for the love of your life? You can create a shopping list that offers up a seven-foot-tall, green-eyed banjo player who loves Paris and is a natty dresser. An online service will probe you and anyone else who joins, subject that information to some sort of algorithm, and give you a list of compatibles within any mile radius you choose.

This process is far more accurate, and even safer in certain ways, than picking out the least slobbery barfly at the corner tavern. However, there are also risks — and good reasons to follow certain guidelines before casting your net into the sea.

First, some online dating services require you to trust them before you know anything about them. Before you can even see the home page or explore the company information, let alone view people's faces, you may have to join and provide your email address, zip code, and birth date, all of which are personally identifiable pieces of information. You may even have to fill out a considerable part or all of a profile before you can access the site.

Match.com, eHarmony.com, and jdate.com (for Jewish singles) are some of the exceptions. You can access their home pages, meander around in the company-related information, and learn the costs before signing up.

(Match makes it *look* like you have to sign up first, because the background is grayed and we computer users have been taught that this means those links won't work. However, the faint, tiny links at the bottom of the page do actually work, and you can read their Terms and Conditions and Privacy Policy.)

Second, know that any dating service will be collecting a *lot* of personal information. This makes sense; the more accurate you are about your own characteristics and interests, the better your matches will be. However, taken together, you will be offering enough personally identifiable information for their companies to know exactly who you are, what you like, and what you might be coaxed into buying. You'll also be telling your yearly income, and some companies won't give you a choice about not answering. You won't be able to continue to the next page of the form unless you do. Realize that you are giving them a blueprint of what products you use and buy, and your buying power.

These sites ask for many bits of information that will identify you, but that they don't actually need to help you find the love of your life. For example, they don't really need to know your birthday, just your birth year, so that you can be classified into an age range. They don't really need to know where you were born. For matching purposes, the region would do. They don't need to know the high school you graduated from. But taken together, this information could let some computer figure out most of the numbers of your social security number. (And certain answers are the same answers needed for the security questions you set up on your online bank accounts.)

Such questions are salted throughout the form, nested in lists of innocuous queries. By the time you reach them, you've gotten used to answering, and you've already filled out pages and pages. I imagine most people are tempted to say, "Oh heck, I've invested so much effort already, I might as well just finish the form. Maybe it's almost over." (But doesn't it bother you that these sites are so slick about information extraction? It bothers me.)

Each dating service has to reveal what it will do with your information, and it is stated in the privacy policy or terms and conditions. Most sites use various parts of your information to build their own businesses. For example, the match.com privacy policy states that it is a part of a family of businesses (IAC) that includes ask.com, citysearch.com, and other online businesses. (You may recall from chapter 13 that ask.com automatically installs multiple cookies on your computer.) You have to go to the iac.com site to find out that one of their brands (online services) is dictionary.com, rated by the *Wall Street Journal* as the highest installer of tracking devices.

Dating sites collect and store your personally identifiable information — your computer's IP address and your interests, activities, gender, age, and demographics — and obtain information about you from other businesses in their group, as well as from business partners, contractors, and third parties. They also collect and store personal information you provide about other people. For example, if you send your mother shoes from one of the sister brands owned by the same company, they will store your mother's name, address, and phone number.

Many sites automatically collect the content of any undeleted cookies that your browser previously accepted, as well as other web pages you visit and the links you click. They use web beacons to track your use of the site and include them in email messages they send to you to determine whether those messages get opened and acted upon. They also allow third parties to track and collect information from their members.

Thus, by using most dating sites, you enter tracking mania. Your every move will be tracked and, very likely, shared with a wide commercial audience.

Of course, some cookies are necessary for the site to operate, but most of that tracking is not for your protection; it's for commercial purposes, so that various partners or affiliates can present you with targeted ads.

Take a look at the "privacy policy" at spark.com (as of September 2012):

> You agree that we may use personally identifiable information about you to improve our marketing and promotional efforts....
>
> You agree that we may use your information to contact you and to deliver information to you that, in some cases, is targeted to your interests, such as targeted banner advertisements, administrative notices, product offerings, and communications relevant to your use of the Websites. By accepting this Privacy Policy, you expressly agree to receive this information....
>
> 4. Our Disclosure of Your Information
>
> Due to the regulatory environment in which we operate, we cannot ensure that all of your private communications and other personally identifiable information will never be disclosed in ways not otherwise described in this Privacy Policy. [Read: "It's not our fault."] By way of example (without limiting any of the foregoing), we may be required

> to disclose information to the government, law enforcement agencies or third parties. Under certain circumstances, third parties may unlawfully intercept or access transmissions or private communications, or members may abuse or misuse your information that they collect from our Websites. Accordingly, although we use industry standard practices to protect your privacy, *we do not promise, and you should not expect, that your personally identifiable information or private communications will always remain private.* [Italics added.]

Privacy policy? Dating sites would more accurately call this their antiprivacy policy.

The most restrained site in this regard appears to be eHarmony.com. The privacy policy is very clear, and the company appears to limit and withhold most of your personally identifiable information from web beacons that transmit to outside parties. Also, they make it obvious that you can explore their site before you join.

(And remember that, as mentioned in a previous chapter, you can opt out of third-party ad targeting by going to www.networkadvertising.org.)

A Working Assumption

By signing up, you are guaranteeing that you are single, that you haven't been convicted of a felony, and that you are not required to register as a sex offender.

Of course, if you have been convicted of a felony, you might not consider it a big deal to lie and sign up anyway. Likewise, a sex offender might not be an entirely honorable person. He may join, even though he is misrepresenting himself.

Most dating sites *do not* conduct criminal background checks on its members at this time, nor do they necessarily verify member statements, although they do reserve the right to do criminal checks in the future. eHarmony.com does monitor activity and investigate complaints, but it does not conduct criminal background checks.

Safety Tips

The safety tips found on match.com and eHarmony.com are good ones. Read those entire pages before you complete or post your profile, and make corrections if you've exposed too much. (Links to safety tips, privacy policies, and terms of use are at the bottom of the page for most sites.)

Choose a site that allows you to hide your profile until you've had a chance to peruse the site and get a feel for the types of people who join. Choose one that allows you to communicate through an anonymous email network on-site, that lets you choose when to reveal your identity, and that screens profiles and photos for appropriateness before posting. Match.com has safety videos that illustrate common scams.

As you fill out your profile, you may be asked for some very personal information that you might be uncomfortable sharing. You might not want to reveal your degree of satisfaction with your own physical appearance or give details about your physical characteristics. And you may have to answer those questions in order to complete the profile.

This is a boundary violation: asking sensitive questions without giving you a choice to not answer, when you don't even know what kind of people join (or own) the site, knowing they will have that information forever. So if you get very uncomfortable with a set of questions, don't answer. If the form will not work without those answers, exit and protest to the company. Ask if they can enable the form even if you don't answer those questions.

Your Fellow Members

Members of online dating sites fall into five categories — men and women seeking:

- Marriage or commitment
- Companions
- Dates
- Faces (some people use the site the way you might use a catalogue; they like looking at the pictures but don't plan to interact)
- Something else (perhaps something dangerous or abusive)

Certain sites lean toward commitment and marriage, such as eHarmony.com and match.com. Others, like chemistry.com and spark.com, place more emphasis on dating. Start with a site that fits with your goals, or with match.com, which has the largest database, even for ethnic and religious groups.

Most sites allow you to become a member at no cost, and you might be able to search profiles, although most require you to complete your own profile before you can see anyone else's. But to see who matches you and to communicate with other members, you must subscribe. When your subscription period comes to an end, you will probably be renewed and charged automatically until you cancel.

One exception is PlentyOfFish (pof.com), which is free. The disadvantage to this site is that attractive women can be harassed with numerous daily emails and even hate mail if guys don't get a response.

To compare the various programs, enter "online dating comparison" into a search engine.

Representing Yourself

Pick a screen name that does not identify you and that is not (even unintentionally) sexually suggestive. You may think Bedlover is about your mattress business, but most people will get a different idea. TeacherSueNSchenectady provides too much information. You want to maintain control over whether another person learns who you are and where you live.

The caveat in the previous chapter — about discovering the ramifications of the impression you try to project — applies even more when searching for a companion. After you fill out the questionnaire on a dating site, before you save or post it, look over it one last time in terms of your need to be seen in a certain way. Are you revealing more than you should out of that need? What would a stranger see in what you've reported? Would you be wise to indicate some of your boundaries so that you are less likely to attract a person who will use, abuse, or manipulate you?

Think about this same issue in terms of the picture you use. If you are looking for a man, remember that men are visual creatures. Put only the most important points you want to make in your text. Most will pay more attention to your picture than your words, except for finding something to reply to. Don't use a picture that is too revealing, unless all you're looking for is a lot of men who want to bed you.

In contrast, most women will read every word. If you're looking for a woman, write lots of words. Describe yourself, your interests, what makes you unique, something about your family, and anything else you feel is relevant.

Whatever your gender, use a picture taken within the past few years in which you look much as you do now. The best photos aren't posed and show you doing something you like to do. Match.com has good suggestions about selecting and uploading pictures.

Guidelines

Be honest about who you are. Don't make up things about yourself. If you are looking for a lasting relationship — even a friendship

— any misrepresentation will be discovered. The more you know about yourself and what you want, the better.

Don't write about what you *don't* want. Instead, flip it over and define what you do want. Don't say, "No beards." Instead, write, "Clean-shaven."

Know your deal-breakers. If you absolutely, positively will not live with someone who has a mammoth dog, be clear about that. But do weigh the importance of your deal-breakers. If your otherwise perfect match has a mammoth dog, maybe you can make room for Yippee.

However, if your deal-breaker is that someone must be clean and sober for a minimum of five years, and someone who contacts you just started recovery meetings last week (after reading your profile), walk away.

To get perspective on your deal-breakers, talk with friends who are already in successful relationships.

Interpreting Profiles

As you read profiles, work toward finding a balance. When you compare someone's profile to your list of what you want, discriminate between the most important items and those wants that are negotiable. A mismatch in politics might be less important than a match in your primary values.

Women may say they're adventurous because they think that's what men like. If you want a companion while you explore jungles or a remote wilderness, and a woman you're drawn to says she's adventurous, be sure to ask her what she means by that. She could mean she's okay with a retro roadside inn with a black-and-white TV.

A man may emphasize his romantic nature because he thinks that's what women like. When you first meet, see if his online story was a memoir or a fantasy.

Someone who insists on a partner with a slender or athletic and toned physique may not stick around for the long haul as a person ages and their body changes.

Addicts

Addicts are a special category. If you are an addict and you're looking for a fellow user, you're in luck; you have a large population to choose from. (Not implying that dating sites are rife with addicts, just that a lot of people are addicted to *something*.)

For that reason, if you are in early recovery from an addiction, quit your membership in any dating site. It's too early. You could easily stumble onto someone who will lead you back into trouble. It takes a couple of years of ongoing recovery to begin to have good judgment.

If you've been in long, steady recovery from an addiction for three or more years, you probably know a lot about addiction and addicts, and you have a good support group to help you vet potential partners.

Don't know much about addicts and addiction? Here's a brief course:

- Addicts will lie to get what they want. If they want you, they will misrepresent themselves to get you.
- The substance or addictive activity will always have more power over an addict than you will. It's not a matter of not loving you enough; the need for chemical stimulation in the brain will override even an honorable person's values (even if the addictive activity doesn't involve ingestion of a chemical — for example, addictive online gaming, gambling, or stalking).
- If an addict says he is in recovery, ask how long and what type of support he uses. Any of the following answers is a red flag.

"I don't go to meetings because I:

a. can do it myself.
b. don't believe in all that God crap.
c. am not like other people.
d. don't have that bad an addiction.
e. can quit anytime.
f. will quit for you."

Such comments are common among people in denial about the extent of their addiction. Such a person is not in recovery. Even if he isn't using this particular week, he is still practicing his addiction.

If he's been in recovery less than a year, he shouldn't be dating, anyway. Newly recovering addicts are prone to handling withdrawal by getting into a relationship too soon. The subsequent drama helps those thirsty brain cells.

A person in good recovery is faithful to his meetings, works the program, and uses support.

Testing a Possible Match by Email and Phone

Once you've opened communication with a potential match, exchanging emails will reveal more that can be useful in ruling someone in or out. Someone who asks, "Do you shave your armpits?" or a Ted who says, "Alice, do you realize that if we date we'll be T and A?" (*T and A* being guy code for "tits and ass"), would be ruled out right away.

What's wrong with those questions? They reveal that this person doesn't understand safe and respectful boundaries for a beginning relationship.

On the one hand, ignorance about such limits is understandable. The rising generation of adults has entered maturity since the advent of Facebook and Twitter. They have never known a world in which people weren't broadcasting their private thoughts and actions. If they grew up in a family in which healthy boundaries weren't demonstrated and discussed, an instinct for appropriate limits may not have developed naturally.

On the other hand, wouldn't you prefer a partner who *did* grow up in a family that promoted boundary intelligence?

So, as you are evaluating email sent to you by potential dates, be looking for signs that they understand the range of appropriate behavior for someone they don't really know. If someone displays boundary ignorance, let them go — unless you are interested in becoming that person's trainer.

What if someone emails you a picture of her breasts? Is this good news or a bad sign? It's another boundary problem and a red flag. A person who exhibits sexual behavior outside the normal context for a beginning relationship has a problem — probably a personality disorder. Drop this person like a hot turnip.

Move fairly quickly from emails into phone conversations. Notice whether the other person listens and reveals at about the same level you do. Do you like their voice, the velocity of their thoughts and words? Do you feel superior, inferior, or equal? Is their use of grammar similar to yours? Do you like their accent? Do they talk too long, in monologues? Are their answers too cryptic? Do you look forward to more calls, or do you dread them?

Do not rely on texting to know the other person. Letters and conversations give you much more information. A terrible recent scam involves creating a false identity, including a false picture, and having a predominantly Facebook and/or texting relationship. The innocent person gets more and more involved, developing feelings for a deliberately constructed perfect match, and then crashes when either the scam is revealed or the non-person allegedly dies in some upsetting way.

Emailing and phoning only go so far in revealing a person. It's easy for people to showcase their strengths and forget about (or even be unaware of) their own warts. This is another reason why it's unwise to reveal too much of yourself before you meet the other person and can begin to see their true nature.

Plus, in person, the other person's energy might be quite different than you imagined. They might be very good on paper yet generate no connection, no spark, when you meet.

Meeting in Person

Until you have a very good sense of the other person, use the following cautions for the first several dates with anyone, *no exceptions*:

- Do not reveal your home phone number.
- Only generally reveal where you live. For example, if you live in a huge city, you might say, "North of the capital building," or "South Hong Kong."
- Keep your private information private.
- Don't give your last name.
- Meet at a place and in an area where you don't usually go, nowhere near where you live.
- Start with coffee or tea in a populated place in daylight.

If you make a personal rule to follow these guidelines to the letter, *with no exceptions*, and your new match tries to talk you out of following them (even if they have a good excuse — for example, "I'm only in town for a few more days"), drop this person. This is a sign of several personality disorders. There's nothing but trouble ahead.

For a first date with anyone, arrange in advance for a friend to call you on your cell phone fifteen minutes into the date. If necessary, this can be your escape call. "That was my friend. I need to go," gives you a graceful way out, even though everyone understands what's going on. It's okay to leave or stop the date in this way even if the person is nice or okay but clearly wrong for you.

If the person seems weird or frightening or potentially dangerous, don't say, "You aren't for me," or challenge them or explain. Simply use your escape call to leave. If you have any bad feeling at all, leave, even if you can't explain why. And if the feeling is immediate and strong, leave before the escape call. As soon as possible, block this person from your profile on the dating site. Also block their email.

Women: Arrive very early, and park in an area that can't be viewed from the meeting place. Leave after your date, so you can see their car. Check to see that you aren't followed when you leave.

During the date, notice if the other person acts in a way that fits the setting. Are they too loud? Are they behaving in a way that makes the people in that environment uncomfortable? Are you uncomfortable? You want a person who is aware of where they are, someone who understands contextual boundaries.

Does it seem easy to talk with them? Do you have to pull teeth to get them to participate in the conversation? If you ask a question, do you get a full response or a curt answer, as if they feel like they're being interrogated? (Remember Blocked Bart?) However hard you have to work to get something going at the beginning, that's a fraction of how hard you will have to work to sustain the relationship if you continue with this person.

Is she a Leaky Lucy? Does she say things that would be more appropriate at a later stage of the relationship? Does she tell you more about her exes or other dates than about herself? (And what do you imagine the headline will be when this Leaky Lucy describes you to that next date?)

Is he a Happy Helpless? Does he take pleasure in being victimized? "I went to the information booth, and they didn't know anything, and then I asked a mall cop and he had just started working here, and the clerk at DVDs R Us didn't even know this coffee shop was in the mall — so that's why I'm late. Sorry." Translation:

"It wasn't my fault. I'm not responsible for being late. I'm helpless among all these ignorant people. I don't know how to read a map. You shouldn't expect me to leave early enough to take these factors into account."

Fast-forward two years. Guess whose name will be inserted into sentences like these? "______________ didn't set my GPS." "______________ didn't tell me you were coming for our usual weekly lunch until this morning." "______________ used up all my gas and didn't refill the tank."

Does he both listen and offer information? Are you truly conversing, or are you having serial monologues in which each response has no connection with what was just shared?

Are you able to talk with each other about a variety of things? Someone who can only talk about types of resin for violin bows — and who knew there were so many types? — may demonstrate one-sidedness.

Is he consistently negative and judgmental? Does she say the right things but behave contrary to her words? Are you being heard, or has he twice asked a question that you already answered six minutes ago? Does she contradict or argue with every statement you make?

Notice if you are with someone who can make room for who you are. Notice if you are feeling increasingly relaxed or increasingly tense.

Layla met Anders at a Starbucks. He bounded in, long and lanky; kicked out a chair and sat; then pulled himself close to the table. He noticed her sweatshirt, the Olympic rings over a British flag. "Did you go to the Olympics?" he asked, his tone even, sounding interested.

Her face lit up. "I wish. I wanted to go so bad. I love the Olympics. It would have been wonderful to be there."

"I don't," he responded. "I don't see the point."

Putting aside a blindness to the meaning of thousands of athletes sacrificing many of their early years and normal kid activities to grueling training, injuries, and medical repairs, or of a biennial armistice in which the peoples of many nations join to celebrate the pursuit of excellence even when they are enemies — this is a serious indicator of a problem.

Can you identify the issues here? Is his comment a:

- setup?
- put-down?
- lack of attunement?
- indicator of good mate potential?

First, he set her up. His question sounded like interest and invited her to show something of herself.

Second, he was not attuned to her energy and excitement. If he saw that she lit up, he either didn't take it in, didn't care, or had no interest in making room for her experience, despite his difference in perspective.

Third, his comment shut her down and could make her feel wrong. At the least, it could reduce her willingness to share her life joys.

Layla went cold. She felt as if she had been slapped in the face. She was so shocked she couldn't sort out what was wrong with his response, but she trusted the twist in her gut. She stood, gathered her things together and said, "Excuse me, I need to find the ladies' room." And then she swiftly left, turning out of his sight as quickly as possible and using a route through a department store to return to her car. Then she drove to another busy mall and sat watching, just to be sure he hadn't followed her.

Now, there's no indicator that he was actually dangerous, but why take a chance? As she watched, she called a friend to sort out what was wrong with what he said, and also to get care for the hurt from being so brutally slapped down.

This was her first and last date with Anders. Eventually, Layla understood that had she pursued a relationship with Anders, she would have been in an ongoing struggle to hold her own ground.

If your gut reacts negatively to the person you meet, leave. You can use your escape call or the bathroom excuse, or just go. You do not have to be polite. You are always allowed to protect your safety first, before anything else, and that includes the safety of your psyche, your tender inner being.

You Click

It's wonderful when you both feel that spark. Still, you hardly know each other. How each of you proceeds will tell you a lot about the kind of boundaries the other person will or will not respect.

Wait to reveal the more sensitive or vulnerable things about yourself until you have more of a sense of who the other person is. More personal information, such as the fact that you were abused as a child or your mother turned tricks, should remain private until you've earned some trust with each other.

Do reveal, fairly soon, if you've had a criminal past, how you feel about that, and what programs you actively use to keep yourself on the right side of the law. (And do use them.)

Get to know each other before getting physically intimate. Go at the pace of the slower person. Trust your own speed. Don't make yourself go faster than you are comfortable going. If you feel pressured, it's a bad sign. If you find yourself giving in to more intimacy than you are ready for, it's worse still. Get out of the relationship.

Notice whether or not it is easy to be up front about your limits. If it isn't easy, is it because your default pattern has kicked in? Or is there something about the other person that makes it hard? Pay attention to this. If it's hard to express what you want now, when your date is on good behavior, it will be even harder further into the relationship.

If the other person goes past your limit — "You are so beautiful I couldn't stop" — you have been given two very important messages: He will make his transgressions your fault, and he will let his desires supersede your limits.

Staying Safe

If, as you get to know someone, you realize that she has misrepresented herself, end the relationship immediately. Don't buy her fervent promises to do better. "I wanted you so much I was afraid you would reject me if you knew this" only means that she finds it acceptable to lie to get what she wants.

If someone has a criminal past, what matters is whether they are actively involved in programs that will help them lead an honest life. For example, if someone stole money because of an addiction, are they in recovery? Do they work an active program? How long have they been in recovery? Go to an open meeting with them and notice if they seem known and welcomed.

Trust your gut. If something seems off at any point in the process, stop right where you are and back out. You can block a person from following your profile.

Do not make excuses for someone. In fact, if you find yourself making excuses — to either yourself or others — get out. If you don't, you'll be making excuses for that person the rest of your life.

EXERCISE: Pop Quiz

Here's an actual interchange. Women, what would you do?

My relative, living in Colorado, called herself Girl of the Peaks and posted numerous pictures of herself in a variety of settings. In only one of these, showing her hiking a level path (no background hills) on a hot day, she was wearing a tank top. One potential suitor wrote to her, "Are those the mountains you're referring to?"

What would you do?

(She ditched him right away. I hope you would, too.)

Ending It

If you choose to end a relationship because the person seems scary, off, weird, or dangerous, you get to disappear. You owe that person nothing. It's more important to be safe than to be courteous.

However, if you are ending it because you just didn't feel what you wanted to feel, you can tell it won't go where you want to go, or you seem too different in some important ways, do drop a brief note saying good-bye.

If someone says good-bye to you — or disappears — accept that as the end of the exploration. Do not persist or chase the person.

Online Don'ts

- Don't complain.
- If you are a blogger and want to describe your terrible date, don't say, "Alfred Example is a jerk." Don't state the name of the loser. Stick with his behavior, your story, and your own experience. For example, "This date played with his Talking Tom app through the whole meal. I felt like the wallpaper."
- Don't curse, not even using acronyms like BS or WTF, as you're describing a fiasco.
- Don't reveal negative or private information about another person (remember, they can easily do the same thing to you).

- Don't bash exes.
- Don't bash previous dates.
- If a former date bashes you online, don't respond in any way. Let it go. (Don't Unfriend the person in Facebook right away; this way, you can see anything they might say about you. Wait a couple of weeks.)

Do report to the police anyone who stalks you, has lied about their criminal status, or is a sex offender. Also report such a person to any dating site you belong to.

Fantasy Island

Cyberspace is a fantasyland. It's a name that conjures the image of weightless star travel and makes mind-to-mind relationships possible. But these relationships are inherently limited. Our closest and deepest relationships, sexual and otherwise, require living contact.

We are already prone to fantasy when we imagine falling in love or finding that special person. Add that pursuit to the bodiless exploration of cyberspace, and our distance from reality can be quadrupled.

Instead, we need to keep our feet on Mother Earth.

Enjoy the stars in your eyes, but use clear sight as you pursue a special relationship, and ground yourself by following a consistent, unyielding self-protective routine.

Chapter 17

THE ART OF INTIMACY

What is the purpose of phones, texting, tweets, or using Facebook? Setting aside the organizational support given by a smartphone, what is the goal of using these devices?

Communication, obviously. But I propose that there's a higher purpose: connection. I propose that we want to keep in touch with people we care about. We want to know what they are doing and thinking. We want to know about their concerns and celebrations. And we want to share ours with them. This is a desire for connection, for intimacy.

Since the advent of all our devices, are we more connected with the people in our lives, or less? I'd say we're connected with more people, including people who would be on the periphery of our lives if we wrote paper letters and put them in metal mailboxes —

people we'd talk to a couple of times a year if we had only one phone plugged into a wall at home. Instead, those people can be in contact with us, and us with them, multiple times a week.

But each of these contacts takes time.

Where is the time coming from? It's pulled from family dinners, walks in the park, and intimate phone conversations, due to frequent momentary interruptions by a ringer or a vibration.

Nearly every potentially intimate moment with the person you are with can be disrupted, and this changes the dynamic. When you know that your heart-to-heart talk can be interrupted, it adds a dimension of tentativeness, of suspense. You don't know exactly when an interruption will occur, but it's likely, so you might hold yourself back a bit, not wanting to get into a deep part of yourself and then go on hold. Even if the disturbance is as brief as the other person reading a short text or checking to see who's calling, it is still a halt in the flow that was building between you.

Intimacy requires a protected space. I fear that a couple of generations have already entered adulthood primed for interruption, and that this is keeping them from accessing and gaining the skills for deeper union with fewer, but more significant, friends and mates.

Once upon a time, teenagers spent long hours on the phone with a couple of their friends and the person they were going steady with. This bothered some parents, but therapists viewed it as part of adolescents' normal development of the capacity for intimacy.

A teenage girl could pour her heart out to her best friend on the other end of the line and feel very close, because her friend's voice was in her ear. At the same time, because the two people were in separate locations, it wasn't too intimate. Either teen could regulate the closeness by finding a reason to hang up. After some internal regrouping — by doing homework or chores, talking to Mom, or playing with the cat — either teen could call back and reconnect until she reached her next limit of relationship tolerance.

This was a transitional phase on the way to intimacy — an

important building block that helped teenagers develop the capacity for vital and deep connections. If, these days, communications are frequently brief or interrupted, where is that development taking place?

Today, texting, tweeting, emailing, and posting on a wall do offer a way out when one's tolerance for relationship is reached — you can ignore a text or respond to an email weeks later — but what is offering a way back in? These truncated or public communications are not meant to be an end point.

Is the plethora of condensed, broadcast, or one-sided messages creating a generation or a culture of people stuck at the level of adolescent development? I think there is some danger of being held at a level of connection that is meant to be just a section of the path on the way to the incredible experience of union that is true intimacy. I wonder if the fascination with celebrity lives or with marriages that result from a televised contest is a way of filling the emptiness of not achieving a deep bond, either with one's own self or with another, proximate person.

Friendship Once Removed

Astrid and Lana were once close friends. About once a year, they'd have a colossal misunderstanding with each other, disentangle their orbits, not talk to each other for many months, and eventually meet and work it out. Then Astrid moved away. When the earth transited through the atmosphere that gave rise to the next colossal misunderstanding, they did not fix it. They stopped all communication.

However, they did not Unfriend each other. Their posts still showed up on each other's pages. After a couple of years, each began responding to the other's posts. Lana would frequently share a post from Astrid. Astrid, in turn, would often Like a posting of Lana's. Here the former friendship sits, and it may stay there.

Whatever the reason for not repairing their rift, Lana and Astrid still care about each other. So this is communicated by this interesting, almost interactive response pattern.

Facebook has inadvertently made this possible. It isn't intimacy, but it is, strangely, a connection.

Marriage

Entering a marriage in which both members have intimacy skills and the ability to attach leads to a lovely, protected world where loneliness is rare and attunement grows. Of course, any marriage will face problems and the partners will disagree, but if they are both skilled and understand how intimacy develops, these can be well handled and even advance the relationship. In contrast, many marriages are do-it-yourself situations, in which each member has partial skills and one or both have an attachment disorder.

The intimacy practice one gets from play, squabbles, and attachment to friends in childhood and young adulthood offers remedial experiences that can make up for parental lack to some extent. But what if young experiences center around digital communication and parallel play? (Parallel play is one of the steps toward social interaction — two people do their own thing in the same area. In terms of social development, it's the first stage of play, occurring in ages two to three.)

The Native American people had a phrase: to the seventh generation. It meant that when they made a decision or considered a situation, they debated about the potential effect on the seventh future generation. Are we considering, alongside our pleasure in our portable offices and colorful interesting games, the effect of such interrupted lives and digitally facilitated friendships on the seventh generation?

Broadening Your Own Abilities

Think of your closest friend. Do you spend more time actually being together or connecting through digital means? When you're not face-to-face with your friend, what is your general communication pattern? Do the two of you mostly talk by phone, or text each other, or send emails? Does one of you lean more toward a certain form of communication than the other? Does one of you initiate contact more often than the other?

Do your current communications with your friend give you the level of intimacy you want? If not, say so, and propose changes for how you connect. Here are some examples:

> "I'd like to talk more and text less."
>
> "Can we talk longer on weekends and text during the week?"
>
> "I love corresponding with you. My days are so busy that I can't usually have a spontaneous chat. My ideal would be for us to send emails and instant messages at night, when my family is asleep, and then schedule a long call once a month."
>
> "I'd like to actually get together more often, rather than just using machines to connect. Are you open to the idea?"

You can ask these same questions of everyone you're close to. This enables you to change how you interact with each person in order to create the level of intimacy you want with them.

If you don't examine how you most want to connect with someone you care about, in a sense you are violating the boundary of your own experience by not creating the opportunity for your preferred degree of intimacy with them.

When it comes to building and maintaining intimacy, each type of communication has different limitations. It's difficult to resolve arguments in a text message. Emails don't provide as much opportunity for negotiating and resolving conflicts on the fly as phone calls. Phone calls don't offer as much opportunity for personal reflection. When reaching out to connect with someone, don't just grab the handiest device; use the form of communication that will best support the kind of relationship, and the degree of intimacy, you want to create.

Your Response to Bids of Intimacy from Others

Look also at the statement you are making by the way you respond to your friends. What is the pattern of your texts' content?

- The logistics of your day?
  - Leaving work
  - Playing poker
- The highlights of your week or month?
  - Was complimented on work project
  - Ran the Prairie marathon
  - Bought a new couch
- Things going wrong?
  - Took 3 hours to get to work
  - Tree fell through roof

Do you respond to your friend's every text, or do you ignore those that don't interest you? Would you prefer to text or email rather than call? If so, what is that about? Is it just to save time, is it to control the dialogue, or is it to avoid too much involvement?

Do you feel more in control of the degree of closeness using a device? Is it more comfortable to text, tweet, or email than to get together in person? When you are with a friend, are you actually *together*, or are either or both of you diluting the experience by interacting with your phones?

If you are both working with apps during your face-to-face meetings, that's called parallel play. It's not intimacy, and it doesn't advance intimacy.

Texts and emails can be a way to fill in the spaces between face-to-face meetings and to stay in touch when time zones or busy schedules preclude frequent phone conversations. However, this is a level of communication that can keep the other person at a distance. Likewise, parallel play limits how much the relationship can advance. When this is a substitute for interaction — when it's used as protection from intimacy — that's something to look at.

A friendship based on texts, email, or parallel play — one that has minimal actual involvement — does not give good practice in how to handle differences, how to negotiate, or how to resolve conflict. If there's an issue, it's too easy to simply disappear until enough time passes for each party to pretend there wasn't a problem, rather than be faced with the necessity of working it out.

If you're not actually with anyone for an extended period of time, there's no opportunity to learn intimacy skills, which include:

- handling your need for quiet time
- making time for solitude without doing something to make the other go away
- managing your retaliatory impulses so that you aren't acting out of anger, which will harm the relationship
- learning how to show up for the other when you don't feel like it

Myriad nuanced skills — attunement, joining, quiet togetherness, trust — that make intimacy possible simply can't be built in a text or email. Electronic communications are too blunt an instrument.

Of course, you get to choose your level of intimacy with the people in your life. However, as you look around or as you look into yourself, if you realize you aren't feeling a deep connection with at least a couple of people in your flesh-and-blood world, consider spending more time in the presence of the people you are most drawn to.

Give yourself practice in being with a friend one-on-one. Put your phones away. Do things with them, or do nothing with them; just be together, walking, talking, and sitting. Pay attention to what you are able to handle and what you don't know how to handle as you are with your friend.

Many therapists are experts on relationships; if you need to build relationship skills, joining a group led by a good therapist is an excellent way to make up for lost time. Certain therapeutic processes, in particular, streamline that development. For more help building intimacy skills, look in appendix C.

Remember, a communication device is a tool. It's the vehicle. It's not the destination. Actually being in a friend's presence, sharing life experiences together, is what advances a relationship.

PART VI

GROUP CYBER-BOUNDARIES

Chapter 18

BOUNDARIES IN CHAT ROOMS

Chat rooms can be a wonderful way to work on a class project, gather periodically with a widespread group of friends, attend a recovery group that fits your odd schedule, or hang out with strangers who share a similar interest. But chat rooms only work well if the people in them maintain respectful boundaries.

Consider this group of friends who stay connected despite the fact that some members have relocated to exotic climes.

"When do you go back to school, Shawn?"

We wait. Time passes. We wait. We wait some more.

"Shawn?"

We're waiting.

"Hello? Shawn?"

We're waiting.

"Next Thursday," Shawn finally answers.

To Shawn, no time has passed. He started reading his email, and when he looked back at the chat room screen, he saw the question and answered it. Meanwhile, the rest of us were held hostage.

By dividing his attention, Shawn weakened a boundary of the group. There's a possibility (isomorphy) that other members will also begin to do Internet tasks while waiting for responses. The integrity of the group will then suffer.

Some chat rooms have the capacity to show the faces of attendees who have a video camera attached to their computers. However, not everyone has a camera, and not every chat room has that capacity. Thus, you may not be able to see some of the members. This leads to two issues: security and etiquette.

Chat Room Etiquette

1. Indicate comings and goings. Say, "Hi," when you arrive, and, "Bye," when you're leaving for good. Create and use a group norm of letting people know when you're leaving temporarily and plan to return.
   - "I'm going for a cup of coffee. Be right back."
   - "Going to pick up my daughter from school. Back in ten."
   - "Need a snack. Gonna fix a sandwich."
   - "I'm back."
2. If you're new to a particular chat, ask if you can join: "Hi, I'm Fireplug. Is this chat open? Can I join?"
3. Don't lurk. If you enter a chat room, say something right away. If you're not sure about the rules of a particular group, you can say, "I'm going to watch until I catch on to how you do things. Is that okay?"
4. If a chat room doesn't automatically display the handles of the people in the chat, give your screen name. Most of the

time, each time you speak, your icon will display along with your words. However, if you have a generic icon, set by the chat server, it may be like others in the chat. In that case, you can either choose an icon so that you are linked with your comments or put your handle in each time: "JED here. I agree with Laser."

5. If you are in a private chat group consisting of friends you know well, you are likely to have leapfrogging conversations and will be remarkably adept at following, for the most part, two or three intertwined conversations.

   An example of such a chat is below. The friends know each other well and have the capacity to insert wordplay and single-word references to previously shared experiences while maintaining the thread of the conversation.

AN EXCERPT FROM A REGULAR CHAT AMONG LONG-STANDING FRIENDS

T: S, did you get ICE there?
S: Is that a rock group?
T: On the rocks???
S: Is it a soft drink?
S: Clueless in Seattle
D: My favorite sleuth
S: Who?
D: Clueless
T: S, Private ear
T: The weather channel was talking about ice storms.
M: I've heard of Iced T...
D: And Mr. T.
M: Music rap group, I think
M: Too gold for me.
T: Tom T
S: Oh, a bit slow here. No, same weather report every day of the year. 50-55, windy and
S: Riany.
T: Or silver
M: Suites me to a...

Continued on next page

AN EXCERPT FROM A REGULAR CHAT AMONG LONG-STANDING FRIENDS (*CONTINUED*)

D: T-shirt
M: Huh? Is that RIANY another Seattle form of rain?
S: No, I was startled when the computer said I'd written too many characters.
T: T-ball
S: Little does it know.
S: I AM writing too many characters.
T: There are only 4 characters here!!
D: And 4 for 1 is great odds
T: Which one is too many?
S: I knew it wouldn't go by.
D: Not me.
S: Actually there are too few characters here.
M: Are we mystery characters?
T: Cartoon?
S: Perhaps, but not fictional.
S: Capitol?
D: There are non-fiction cartoon characters?
T: You mean Pooh isn't real??
M: Pluto is about to be demoted from planet to iceball.
D: Nothing you ever learned is an absolute.
T: I hope Goofy, Mickey, and the others raise a fuss.
S: Pluto water, iceball, pluto on the rocks.
T: Does that mean Alaska is only ½ a state?
S: Most of us are only half a state
D: Part liquid, part solid.
S: Yep, that's us.
T: A state of mind!
T: Does that mean I'm only ½ here? — Wait don't answer that!
S: First Pooh and now this.

Such play is fun for close-knit groups, but when you are in a public chat group with a shifting population, try to stay with one topic at a time so that people don't get lost.

6. If you do want to change the topic, put it forward and see if the group is ready: "Are you guys ready for a change of topic? I wanted to talk more about dingos."
7. Don't be doing something else while you are there. If you

are in the chat room, chat and listen. Don't text, talk on the phone, surf, or read email at the same time. If you do, you will create lengthy pauses that are annoying for the others and slow down the whole conversation.

8. DO NOT USE LOTS OF CAPITAL LETTERS. IT LOOKS LIKE YOU ARE SCREAMING.
9. Don't use lots of colors and art. It's a distraction from the interaction. (Some chat windows automatically separate speakers by the color of their balloon or text.)
10. If you are in a private ongoing group, collect each other's email addresses so that if the need arises (e.g., the chat room's server goes down), you can contact each other to regroup.
11. If you are in an ongoing private group with a fixed membership, make a plan for handling a potential intruder. Have a code word set up that means "everybody go off-line." Then have an alternative room where you can regroup. Set all this up ahead of time so that you don't have to email to find each other, just in case you are discovered because one of you was hacked.

Chat Rooms for Children

I checked out one of the largest kids' chat rooms. To join, I had to enter my birthday. Not knowing what the minimum age requirement was, I entered a false birthday that would make me eleven years old. I was immediately blocked with a screen that said I had to be thirteen. All I had to do was click out and click in again, without even re-entering a new false date, and I was admitted to the room. (I assure you, I am considerably older than thirteen.)

I was warned to follow these rules:

- Age range: thirteen to nineteen
- Don't disturb or flood chats

- Don't give personal information
- Don't hack or exploit the site
- Don't post obscene or vulgar messages
- Use a clean nickname
- Don't exploit the kids
- Only use the site under parental supervision
- Know that, although the site is staffed, the moderator can't be there every minute

I stayed in the chat for just a few minutes. In that short time, two users were flooding the site. Flooding means repeating the same thing over and over, usually using large, bright text. Half the messages were vulgar or obscene. Nicknames being used by participants included Horny Dad, Horny Daddy, Child Model Agent, older guy, cockout4girls, nwtyvids4littlegirls, and tongue.

The ads, which ran across the bottom and side and in pop-up screens, were targeted toward young adult males, ages twenty to thirty-five. Now, why would a site purporting to serve teens have ads targeted to an older population?

In the time I was there, there was no sign of a moderator. No one was kicked off for vulgar speech or a suggestive nickname. I went back five minutes later to see if a moderator had gotten rid of those who were blatantly violating the rules. I found that the content had become even more explicit or violent. (And violent speech is not against the rules.)

I'm guessing participants tested the waters and, when no one showed up to enforce the rules, felt free to be more disgusting.

Do you think parents were paying attention to their children's visit to this site? Were kids keeping to the rule of being supervised while they were there? I doubt it.

Some kids' chat sites are more carefully monitored. For example, Lycos chat provides a great deal more specific information about rules, child protection, and how adults can participate with their children and keep them safe. It also has Navigators — people who can help with a situation.

Members appear to use the site to share opinions, sound off, and learn about specific topics from others. The introductory materials and tips are worth reading, even if your children will visit other chat sites.

Stay Watchful

When you or your children are in a chat room, don't count on every comment being screened. If someone is offensive, you can report it. But with vast numbers joining chats on the Internet, a monitor can't be watching every single room at all times. Online chat services make no promises about their diligence in monitoring users or kicking users off the site, although they do reserve the right to do so.

Therefore, it's up to you to keep your children (and yourself) safe — by not sharing identifying information, by realizing that you will be given cookies that will track you, and by taking your children (and yourself) off-line when a chat room doesn't match your hopes for it.

Chapter 19

WORK BOUNDARIES IN CYBERSPACE

The crux of boundaries at work is: Do your job. Your employer owns your time and energy while you are on the clock. You already know this, but I'll bet you can identify at least one other worker who doesn't seem to.

This means don't roam around the pleasures of the Internet or enjoy digital entertainment on company time. To get your best shot at a promotion, to gain the trust of your manager, for your own self-esteem, and to rise above other employees who abuse this caveat, honor that covenant. Therefore, keep personal calls, texts, and emails to a minimum. Use them only for emergencies or quick check-ins with your child.

If you use the Internet on your job, it can take great discipline to not get distracted by some enticing site. To help yourself maintain this discipline, only open the browser windows you actually need to do your work.

Person vs. Member

There's a great distinction between the two major personas we inhabit throughout our lives. Indeed, many of our errors and much of our heartbreak comes from not understanding the difference between the two.

You are a person. Everything within your individual self constitutes your *person system*. When you are operating from your person system, you have feelings, thoughts, reactions, emotions, memories, goals, needs, and wants that are all related to you personally.

When you join any group, you become a member. You then cross the boundary from your person system to that *member system*. You take with you your energy, capacities, gifts, creativity, and learning edges, and you join with the others in that group. (To get actual practice at making the transition from your person system to a member system and back again, and to increase your facility in determining which you are inhabiting at any particular time, attend a Systems-Centered® Training, or SCT®, workshop; see www.systemscentered.com.)

A simple example of person system versus member system happens when you become half of a couple. While you are riding home from work, you're thinking about your day or the ball game you want to watch later. You are operating from your person system. Then you remember that your mate had a hard challenge at her job today and will be exhausted — and probably late getting home. So you decide to stop by Mamma Roma's and pick up a vat of their fabulous lasagna so that dinner will be easy. You just crossed from person to member.

You made the transition from thinking *I* to thinking *we*.

Good parents learn quickly how to switch automatically from

person system to member system in relation to their children. Despite their involvement in their own interests, such as a focus on the evening news or catching a lost stitch while knitting, they can move out of that personal focus and hand off parental chores like tossing a basketball back and forth between teammates. "Can you pick up Via?" "Yes, can you drop off Le's prescription?" "Yes."

In both examples, the transition was automatic, unconscious, and helpful. At other times, though, not being conscious of the system you're operating from can cause a problem, especially if you're acting from the wrong system for the situation.

We all make mistakes with this in flesh-and-blood life. In disembodied cyberspace, where we're more prone to fantasy, it's even harder to keep the two separate.

Furthermore, tracking that distinction and staying clear about the member boundary can be much more difficult in loaded or intense situations. Many of the arguments that occur in a marriage arise from not crossing from person system to member system. Examples:

"You always choose the movie!" (Translation: "You are always operating from your person system.")

"That's because you never say what movie you want to go to!" (Translation: "You don't show up from your member system, either.")

Within your person system, you know what movies you want to see. To operate as a member, you bring your knowledge of your wants across the boundary and present your list to your mate. Then your mate, crossing from person to member, says what is on her list. Now, two movie lists are in the group. If the same movie is on each list, you have a winner. Otherwise, negotiation ensues, with one member presenting a proposal and the other member countering.

"I propose that we see *Down in the Air*, because it is at the cheap theatre and it's almost out of circulation."

"That's true. It is about to go to video [indicating, as a member, that you are listening to and considering the other's ideas], and my thought is that seeing it on the big screen is not important. However, *Deep Forest* is also near the end, and I think it will be better to see it on the big screen."

"That's a good point. I'm willing to see *Deep Forest*, and I'd like then to choose the ethnicity of our meal beforehand."

"That's fair. We go to the movie I care about, and we eat the food you're in the mood for. Okay."

"Deal."

If married persons keep going after what they want, without communicating and negotiating about those wants, they are not crossing into their member systems. This is how marriages get into trouble. In a marriage, you become a member when you tell your partner what you want and then listen to what your spouse wants. That starts a member discussion.

Accusations and recriminations between married persons are usually arguments from one person system to another person system. When they each learn how to cross into the member system, the entire tone of the household changes.

It takes a conscious switch of perspective to cross the boundary from your person system into the member system, so that instead of looking out of your own eyes, concerned just with your own needs, you are looking out of a member's eyes, concerned with the well-being of the unit that is made up of all its members.

Entering the work space is also an occasion of crossing the boundary into a member system. From your person system, you bring your talents, ideas, dedication, energy, time, experience, education, and honor. Then, within the member system, you apply all of these to your job and the organization that employs you.

You may already do this unconsciously. When you enter your workplace, even if it's your own home office, you may automatically click into member system. If so, notice where it happens: in the car or subway, as you walk through the door, when you sit at your desk? Notice how it feels, and how your perspective changes from *me* to *employee* or *professional.* Notice your changed outlook, goals, thoughts, and plans.

If you work at home, do you truly switch to member, to your membership in the work effort you are sharing with somebody — clients, customers, colleagues in other home offices, students, or the company that receives your product — or are you divided between your work requirements and the dog, a sick child, or the kitchen that needs cleaning?

How does it feel when you're torn between two demands? Would it feel saner to put a boundary around the work part of your day, to put a Do Not Disturb sign on your door, to have only and all work items in your home office? (I wear a particular cap and put a sign over my doorknob. The sign is for everybody else; the cap is a tactile reminder that I'm on the job.)

It could also be interesting to notice when you change back to your person system after you leave work. When do you pick yourself back up, returning to personal thoughts, wants, ideas, and interests?

Some people stay stuck in their work-member system, and it can make their lives unbalanced. And some people carry their work-member system into their homes and forget to make the transition into their marriage-member system. When this happens, they are still looking through professional eyes and thinking from a professional perspective as they encounter their spouse and kids. This can create many problems, but that's another book.

Guidelines for Ethical Boundaries at Work

1. Cross the Boundary from Your Person System to Your Work-Member System.

If, when you enter your work space, you automatically cross the member boundary and enter into your work-member system, then you are probably already handling online work issues well.

If not, then this is your first goal: to step from personal concerns into the uniform of your work mentality. Think about your job, the niche you fill, your contribution to the company. Think of yourself as a member. Remind yourself of the goal of the organization and the purpose of your job. Give your energy to it.

Can you feel the difference?

Throughout the workday, if you catch yourself slipping back into your personal perspective, notice that difference. How does it feel to be in your person system at work? Contrast that to the feeling of being in your work-member system at work.

Email, Forwards, and Solicitations

You already know that your emails or forwards should relate to your work or your company and be for the purpose of advancing your work effort. Occasional forwards of a joke or a cartoon may seem like a way to lighten the atmosphere in your work group, but remember that you will be interrupting a colleague's train of thought, so keep this to a minimum. People work at different paces. Your rhythm of taking a break may differ from that of a coworker on the other side of the wall or the world.

If you are a manager or boss, draw a clear boundary around soliciting your workers. Even a passive solicitation, like a pile of your daughter's Girl Scout cookies on the secretary's credenza, goes a bit too far; workers will feel some pressure to make a purchase to stay on your good side.

As a boss, manager, or coach you are in a leader role. Within this role, you are still a member of the organization and should be operating out of a member system. Soliciting employees is an example of being in one's person system instead.

Therefore, for a boss to send employees a political endorsement, a petition, a call to social or political action, or an overt request for a donation — even for a very good cause — is a boundary violation.

The only exception is if the entire company is officially involved — for example, in a United Fund drive — and if each person's contribution is truly anonymous. Employees should be informed of this obligation when they are hired.

Is it okay to solicit coworkers? This is also a touchy area. If someone turns you down, will it affect your work relationship? Do you feel you'd have to buy, or buy into, a colleague's cause to ensure further cooperation?

It's safer to simply put a boundary around all manner of nonwork requests and stay focused on the job.

2. Operate within the Boundaries of Your Profession or Organization.

Every workplace or work situation has its own boundaries. Some are stated, some may not be. You may have been taught some boundaries when you were trained, educated, or licensed. Other boundaries may be implicit.

For example, if you are in the military you've been trained exhaustively on honor, safety, risk management, teamwork, and chain of command, and the boundaries that pertain to each situation. The penalties of violating boundaries are clear and sometimes harsh, so you've become disciplined and you respect those limits.

If you work in an office, honor may be implied. You may not have been told not to steal company property. But you know you shouldn't, right? That violates an honor boundary.

So the boundaries for you to respect when using the Internet or the company's digital property may be both implicit and explicit. You may have been told to be honorable about handling those lists of data, but not told to use email only for work.

Your employer might not mind your using the copy machine to make twenty copies of your daughter's swim award, or using the scanner to upload it so you can email it home and then on to one hundred relatives, but did you ask first? Normally, you should.

Getting conscious about where this line is in your workplace is member behavior. Test yourself, if you'd like.

EXERCISE: Explicit vs. Implicit Work Boundaries

List 1. Explicit Boundaries

Jot down the limits on personal Internet and equipment use that you've learned from your training, state regulations, orientation, boss, and/or company manuals. Possible topics:

Computers
Internet
Printers
Scanners
Work phone
Work laptop or tablet
Apps
Browsing
Games
E-readers
Email
Texts
Videos
IMs
Chat

List 2. Implicit Boundaries

Jot down the limits that you either know instinctively or have observed in the behavior of others, that relate to digital or Internet use. Possible topics:

Computers
Internet
Printers
Scanners
Company phone
Company laptop, tablet, or equivalent
Apps
Browsing
Games
E-readers
Email
Texts
Videos
IMs
Chat

If you are in doubt about any specific boundaries, ask. Crossing a certain line may constitute stealing in your employer's eyes, even if everyone does it.

Remember that your use of your employer's digital devices, and your own online work, may be monitored. You may know what your company monitors, or you may not. Cookies could be planted on your company computer that track everything you do and report your activities back to someone whose job is to examine employee use. It is also easy for an organization to install a background program that lets someone pull up your computer screen on their computer and watch what you are doing on your work computer in real time. Something to consider.

3. Don't Bite the Hand.

Don't use your own personal online networking in a way that would embarrass your employer, boss, or coworkers. You might want to complain about your boss's very flexible interpretation of lunch hour, but don't mention it on Facebook or tweet about it. Talk privately with a friend who doesn't work there.

If you publicly bite the hand that feeds you, you might get bitten yourself — in the behind.

"They Owe Me"

This issue gets more complex when someone works for a company that doesn't treat its employees well. Increasingly, employees are expected to return to the working standards of the nineteenth century — sixty-plus hours a week, Sunday afternoons off. An employee who keeps to contractual time boundaries and leaves at 5:00 on the dot is considered not dedicated.

"You want to spend time with your family? You must not be serious about advancing."

Or you may have so much work that the only way to keep up is to work at home on your own time. Although digital devices increase the speed of communication and information processing, they have also made work much more portable. It's becoming the norm for employees to continue their work at home, putting in significant extra hours off the clock. The increase in speed and efficiency is thus offset by the decrease in personal time.

It can be harder to hold on to your work-member-system context at home, especially if your family needs you to return to your family-member system. Also, the anonymity of working at home can feel similar to the anonymity we have while we're driving. We may vent angry reactions with a quick-slap email — a loss of member-system perspective.

Some bosses and business owners abuse their positions. They vent their frustrations on hapless employees, scream at them, blame employees for consequences they didn't create, or sabotage managers by not giving them the power or autonomy they need.

In these work environments, abuse of company property may well go up. Employees may give themselves unofficial compensation or revenge by sneaking in phone calls, texts, emails, or periods of surfing.

Pay attention to the risk. And notice how it feels to sneak or cheat. The cost to your self-esteem and the complications of sneaking may not be worth it.

We each decide how clean our behavior will be in our work. Once you are clear where your employer's boundary is, then you choose whether to abide by it or to draw your own boundary — either closer to honor, or further from it.

Cover Your [Posterior Anatomy]

What if someone at work is out to get you?

Your best defense is to build a good relationship with your boss, and that includes respecting company boundaries around electronic and digital use. If you see that someone is trying to set you up, document the evidence and present it to your boss.

What if the person who's out to get you *is* your boss? Document the evidence and keep a record of your own clean actions. Make copies of your work product, if possible, or document your online use so that you can show that you are doing your job and not wasting company resources. Print or use a thumb drive to copy your documentation and take it home with you to keep it safe.

Absolutely do not email anyone about this situation from your work computer or text about it from your company phone. *Do not* put any comments about it on your Facebook page, not even marked private. The walls (including your Facebook wall) may have ears.

If you love your job and that company, build a relationship, if possible, with some other bosses, so that others in management positions see your worth. I wish it were true that bad bosses would get found out and ditched, but they are more likely to be promoted. In that case, if you've built a good reputation with your coworkers and other managers, the replacement will be an improvement.

Your Professional Appearance in Cyberspace

We used to find resources in a big, heavy book called the Yellow Pages. When you wanted to find a florist, you looked there.

No more. Today, anyone who totes around an electronic device goes to cyberspace to find a product or a service. I wonder how long it will be before the Yellow Pages become an artifact in the Pre-Digital Age section of the history museum.

A florist is no longer competing with the other two or three shops in the average American town but competes with every florist online. It is much tougher to get noticed, especially if fees and services are similar across companies.

Therefore, if you make or sell widgets, are a professional, or own your own business, you need a website in order to compete. A professional massage therapist, manicurist, voice teacher, band, counselor, dog walker, or plastic surgeon who has a website will have an advantage over competitors who don't.

Why not sign up for some company that offers listing services or power exposure, who will put your name in lights for a mere monthly charge? This might or might not be a good business decision, but before you fill out the blanks for your first free scan of your business presence (possibly offered to draw you into such a company's website), *read the privacy policy.* When you press *Scan*, you will be giving your company's identifiable information, and probably your personally identifiable information as well.

As you'll see in most privacy policies, this information may be disseminated to every sales maven in the free world. Furthermore, by submitting your email address, you are agreeing to receive email advertising and telemarketing, *even if you are on the FTC's "do not call" list.* Oh, these guys are slippery.

From a boundary perspective, you are almost always safer going with a small, local web designer or marketing firm; they are far less likely to sell your company or personal information to someone else.

Your Personal vs. Professional Web Presence

You might be able to have it both ways. You might not. The person-versus-member distinction exists even when you are an independent contractor or own the business yourself.

In your personal life, you may play tennis, love Garfield the cat, use non-organic chemical weed killer on your lawn, and be a radical hot sauce collector. Professionally, you are a consultant to organic food coops. If your clients find out about that weed killer, it could exterminate your business.

We don't want our professionals to have personal lives that are even slightly contradictory. So don't put anything on your personal

Facebook page, website, or blog that might damage your business or undermine your professional image. Ditto for anything you say online where you reveal your identity or your email address. We don't mind if Lady Gaga runs around in a bra (and little else), but we don't want the pope to do it.

This is just as true if you are self-employed. As an independent contractor, you are still a member of a profession. If you are the owner of a tractor repair business, you are a member of the tractor and farm world, and a member of the business owner fraternity. One post of *I'm so freakin' sick of tractors* on someone else's blog could hurt your reputation and your business.

This is about appearance as well as intention. Before you set up your personal Facebook page with pictures showing you romping in the back yard with your thirty pet frogs, think about its ramifications for your job, professional reputation, and career. Your harmless fun might look creepy and could cost you your job as president of the Society for Elevating Amphibian Respect. People just might not understand.

Repairing Your Online Reputation

In your wilder, younger days, you let it all hang out. Now that you are becoming a person of distinction, your earlier cavorts aren't so cute. What's a respectable person to do?

As long as no one is after you, you can clean up your online presence by editing any page you've put up yourself — on Facebook, LinkedIn, and so on. Also, turn off Picture Recognition in any of those sites. (On Facebook, for example, when Picture Recognition is turned on and you put your picture on your Facebook page, all other pictures posted on Facebook — and possibly elsewhere — are scanned to see if you are in them. If you are identified in someone else's photo, it is tagged with your name, and Facebook announces your name associated with that photo.) Last, go back over your history of posts and make private or delete, if possible, any that no longer fit your current image. (Chapter 12 describes how to access privacy settings in order to take these steps.)

Unfortunately, anything you ever posted online could still be out there somewhere, and therefore be retrievable. If there's something very embarrassing or damaging in cyberspace that you fear might turn up someday, your best option is to set up a blog where you talk about those times, the lessons you learned, and how you were able to change. The public is more forgiving of a confession than an exposé, especially if they can learn from your morality tale.

Boundaries on Facebook, Plaxo, LinkedIn, ZoomInfo, and More

A professional networking site will usually require you to commit some of your personally identifiable information before you can get very far into it. But don't just fill out a form automatically. Explore the site's terms and conditions and privacy policy first.

Here are some basics about the most popular of these sites, as of this writing:

With *Facebook*, you can give your professional life a public forum, showcase events or products, and invite traffic through Liking or being Liked. The biggest danger of Facebook is getting distracted by other people's posts and following them like a trail of bread crumbs away from your work.

Plaxo starts as an address book that does what Mac users already have in their various i-devices. If you update an address in your address book on your iMac, iCloud will automatically sync it with your MacBook, iPad, and iPhone. If you don't use iCloud, it is still easy to synchronize via direct connections. Plaxo does the same thing, keeping your addresses in cyberspace (actually a computer in a building somewhere) and, with an upgrade, coordinating them among your various electronic or digital devices. You are also invited to upgrade to a personal assistant, who will suggest address changes and corrections. You can add features, such as a calendar that also can be synchronized. Plaxo is owned by Comcast, so it has a long reach and many "com-patriots."

LinkedIn is a directory that you can use to find or be found by other professionals. As with Facebook, you maintain your own page and conduct your own exploration of the others who have joined. The base of individuals available consists only of those who have joined LinkedIn.

ZoomInfo is the Yellow Pages meets the CIA. The database at Zoom is extensive, with detailed profiles of businesses and business-people. Its source material is gathered not just from the folks who sign up but from web crawlers that feed the "ZoomInfo extraction engine."

Is there a limit to the degree of privacy you can count on if you join or post on a professional networking site? Yes. Each has a privacy policy, but most have so many loopholes and exceptions that just about anybody trying to make a buck could possibly have access to your information.

So what boundaries can you set with such sites? It's a catch-22. You want people to know who you are and what you offer. You can't do this without risk. In the olden days, paper directories limited your exposure to a certain geographic radius. Online, your exposure is worldwide. By maintaining a web presence, you are letting yourself be known — by anybody.

Nevertheless, there are some limits you can set. Chapter 12 deals with protecting your privacy and personal information on networking sites. (The tips for social networking sites apply equally well to professional networking sites.)

Job Hunting

You are now marketing yourself. You are the company you want to show to the world.

ZoomInfo and LinkedIn might be good resources for you, especially if you have a unique skill that is best used in a specialized company. You can create a web profile, find contacts, and discover businesses both large and small. You can contact recruiters, see what businesses are looking for, and find out who's hiring.

You already know that you should tell the truth about your skills, training, and experience. (A lie that gets discovered later could cost you your job, or even your career.) It's okay to put yourself in a favorable light, but misrepresenting yourself is never acceptable — and today, with all the electronic tools at employers' disposal, untruths are more easily discovered than ever. So stay clean.

When job hunting, use the same manners online that you'd use in a face-to-face interview. Respond promptly to calls and emails (using full words and sentences and clear, correct English). Be straightforward. Prepare by knowing your strengths and creating a list of questions for potential employers.

Remember that when you look for a job, your past indiscrete behavior — especially if it's online — can come back to haunt you. Before you begin applying for jobs, peruse your entire online presence, both professional and personal, using the mind-set of a potential employer. Remove anything that reflects badly on you, your profession, or any potential employer or client.

Google your own name, and expect every potential employer to do the same. Follow up each link that appears. If something problematic shows up that you can't remove, you may need to write an attachment to your résumé that admits a past indiscretion, explains what you learned from the experience, demonstrates how you would never make such a mistake again, and documents your increased maturity.

Nearly all digital devices are moments away from turning into public forums. To keep ourselves marketable, we must stay aware of the fact that we are inches from the world stage at all times.

Chapter 20

THE BOUNDARY CONTINUUM

Varying Vigilance

The more I've learned about the dangers of the Internet, the more careful I've become — and the more I've instituted measures to protect my computer and my family. This is important to me.

My mate, who has a different computer (and whose brain works differently), doesn't think about the risks and doesn't take extra measures. We place a different value on protecting our computers and our information. And this is okay.

You might have a similar circumstance in your own household. Each person assigns this information a different degree of importance and acts on it accordingly.

Powering On

When the World Wide Web started, we all plunged in without knowing how it worked. As each door opened, we ran through it, getting ever more deeply involved. It took businesses a while to figure out the commercial possibilities, but once they did, the Internet went from being a colorful playground to the most aggressive marketing tool on the planet.

We are hooked on the Internet's speed and convenience, on phenomenal access to an infinite variety of subjects, and on opportunities to see the best and worst of human nature.

And we have to remember that the Internet is run on immeasurable volumes of incomprehensible code, snicking through millions of computers in hundreds of countries. Every single thing that comes to your computer has, behind the curtain, code. Oz has arrived.

We, therefore, have an ongoing choice to consider: How far are we willing to move along the continuum of setting boundaries in cyberspace?

BOUNDARY CONTINUUM

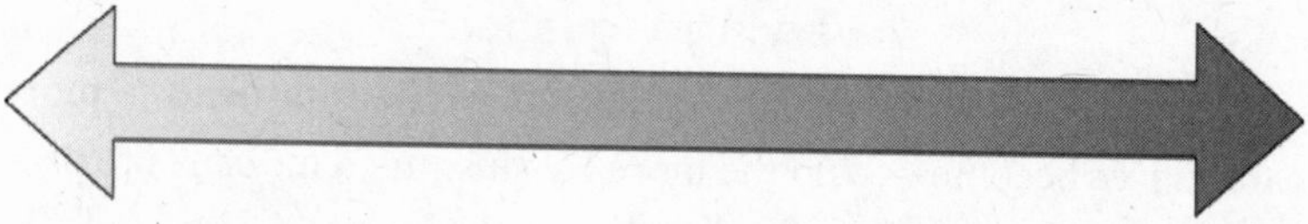

Hold my breath and plunge ⟷ Nonstop monitoring

I started out on the plunge end. I knew nothing about the dangers as they developed. At one point, I had an argument with my brother when he said there was no privacy on the Internet. He turned out to be right. (Don't you hate that?)

As I investigated the variety of ways companies could hijack my information, I would get to a place where my brain felt too big. One day my agent said, "Explain flash cookies."

I thought, *I can't. I don't understand flash cookies. I don't even* want *to understand flash cookies*. I was saturated in techno-speak and had to go cut up fabric for a quilt I was making.

A few days later, I could again approach flash cookies, and they weren't all that hard to understand.

You might discover a similar process for yourself — that you'll reach a limit to what you can absorb and then do. So stop. Come up for air. You can always come back later and learn another bit.

You might end up setting one boundary at a time, in a single aspect of the digital world, and leaving the other digital devices for consideration another day (or week). Go at your own pace; expose yourself only to the amount of information you can absorb. This is, in itself, respectful of your own energy and endurance boundaries.

Whenever you set a boundary — any boundary — you create a cleaner space, within which is you. You make room for your own life. You say to the world — and more importantly to yourself — "I matter."

Here's to all of us mattering.

GRATITUDE

At times our own light... is rekindled by a spark from another person. Each of us has cause to think with deep gratitude of those who have lighted the flame within us.

— Albert Schweitzer

Scott Edelstein, who continues to deserve highest praise, has the most miraculous combination of attributes. He is a scholar and a gentleman. He is wise. He is honorable. He is a very good editor, a brilliant explainer, and a most perceptive moderator. As my agent, he has stood with sword and shield in front of me for over twenty years. Thank you, Scott.

Sherry Ascher takes care of everything around me, protecting my writing space and my writing mind. She has her own version of a sword and shield, also a fierce protector. Thank you, Sherry.

My dearest friends each contribute heart and timely presence. Most have been supporting me in manifold ways for decades. Thank you:

Ann Weston
Barbara Self
Cassie Major
Harry Lynam
Jill and Kevin Shea
Joan and Blaine Haigh
Judy Burns
Karen Selby
Lynn Keat
Mary Zibung
Rabbitt and Cliff Boyer
Shirley Averett

The following friends started out occupying a niche in my life and now have become dear to me. Thank you:

Connie Wolfe
Marge Bagwell
Marilyn Sherman Clay
Rickie Schmadel
Ron Wilkinson
Susan Johnson

Then there are my companions in my other life at Gallery Golf Course — the most beautiful, affordable golf course in the Northwest (and possibly the USA), with a view of the Salish Sea from almost every hole, open to civilians. Thank you, Jim Smith, Wayne Dorrenbacher, Mike Hokansan, Paul Phelps, Lana, Tom, Steve, Bob, and Jim, for including me in your club. Mike Fields, who can drop three strokes off your game with one lesson, thank you for giving me a chance. Denise Conklin is a natural cook, and therefore a talent, who will cook me anything I want and revise it to be abstinent. Thank you, Denise.

To AWF, may blessings surround you wherever you go.

I have such a wonderful family, our bond continually renewing no matter how many miles or months separate us. Thank you:

Robert and Lori Smead, my brother and sister-in-law

Cousins Tom and Susanne Stein

Cousins Jim and Carol Webster

At the end of each movie, the credits are run. Everyone who played a significant part in the production is named, even the dog, cat, or bear.

The production of this book comes to us by the intelligent staff of New World Library, and each person I worked with — my editor, Jason Gardner; my copyeditor, Carol Venolia; and my publicist, Monique Muhlenkamp — was extraordinarily skilled, competent, thorough, attuned, organized, patient, and adult. They all know their business. It was a pleasure to paddle their liquid process.

An author owes accolades to a lifetime of people. A comprehensive list of credits to all the people who keep that pen (or keyboard) moving, and the spirit to return to the work space no matter what happens, would fill a small library. The short version is this: If I've had any kind of encounter with you ever, you've helped. Thank you.

Appendix A

YOUR POSITION WITHIN CULTURAL NORMS

In order to create effective boundaries involving any device or method of communication, it helps to be aware of whether you're ahead of the current cultural norm, behind it, or in step with it.

As hard as it can be to imagine, at one time homes did not have telephones. As people experimented with those newfangled machines, they had to tell someone if they could be reached by phone. "We just got a telephone, so you can call me now. Ed pulled the cash out of his pocket, even though he doesn't think this telephone fad is going to last. He says, 'Who wants to be interrupted all the time?' Of course, he thought having an indoor bathroom was unnecessary, too."

Eventually, the culture slid past the fulcrum, and having a phone became the norm. Then it was necessary to tell someone if you did not have a phone. "We don't have a phone yet. If the time of the potluck supper changes, you'll have to call my neighbor, Elsie. She'll pass the message on to me."

We walk the plank of the teeter-totter in a similar way with each new device. When mobile phones first became widely available, few people had them. No one was expected to own one, and we didn't assume that people would be available by phone at all times and places. Now we expect to be able to reach nearly everyone whenever we call, and someone has to tell us if they don't carry a cell phone.

We are on the far side of the fulcrum with email now, too. We expect people to use email, and if they don't, they get left out a bit or left behind. We also expect everyone to be connected by Wi-Fi or broadband. We have to remember not to send pictures or videos to our few impoverished dial-up friends. (FYI, I'm writing this in early 2013. If you're reading this appendix in 2014 or later, feel free to enjoy a hearty laugh. If it's 2016 or later, find an old person and ask them what "dial-up" means.)

With texting, Facebook, Twitter, and professional networking sites, we are still (as of 2013) on the new side of the fulcrum. We don't automatically assume that everyone uses these, and we don't think that people who choose not to use them are weird or backward.

We are inching toward the fulcrum with texting, however. I recently called a friend and got her voice mail message, which explained that she did not text. So we may be approaching the point where people who don't text feel they have to say so.

Still, just because a device or method is popular, it doesn't mean you have to use it. It just means you'll have to deliberately define your boundary so that you are included if you want to be, and so your friends and family members will know what your digital limits are.

The clearer you can get about what devices and methods you choose to use — and how closely your choices reflect the cultural norms — the easier it will be to communicate with others and to set healthy and helpful communication boundaries.

Now here's another wrinkle. So far, I've been writing about cultural norms in terms of our entire culture. But, at any given time, there often exist multiple, sometimes contradictory, cultural norms — one for each generation. For example, I know of only a few people over the age of eighty who text, and I don't know one teenager who doesn't.

The Generation Profiler below can help. It's an easy-to-use tool that helps you do two things:

First, it enables you to easily assess what devices and forms of communication someone else is likely to be comfortable with, based on their approximate age.

Second, it allows you to quickly compare the devices and communication media you use against the ones used by other people, whatever their age. This will help you communicate with almost anyone in a way that works for both of you.

Take a few minutes to fill out the blank Generation Profiler below. If members of a particular generation generally use the form of communication noted, mark it with an X. If not, leave it blank. There are no perfect answers, and you will probably know of a few exceptions for each generation. That's okay. Just make your best general assessment for each generation.

Note that your own Generation Profiler will probably be very different from that of someone who is five to ten (or more) years younger or older than you.

YOUR GENERATION PROFILER						
DEVICE OR MEDIUM	You use it	Your gen* uses it	Your parents' gen* uses it	Your grand-parents' gen* uses it	Your children's gen* uses it	Your grand-children's gen* uses it
Cell phones (for talking)						
Smartphones (for talking)						
Smartphones (for taking and sending photos)						
Smartphones (for browsing Internet)						
Texting						
Skype						
Voice mail						
Email						
The web						
Facebook						
Other social networking sites						
Professional networking sites						
Twitter						
Blogs						
E-cards						
Physical greeting cards						
Snail mail letters						
Snail mail postcards						
Printed digital photos						
Photos sent as digital files						

* Gen = generation

To get a better sense of how to use the Generation Profiler, imagine that thirty-five-year-old Lee fills it out.

LEE'S GENERATION PROFILER						
DEVICE OR MEDIUM	You use it	Your gen uses it	Your parents' gen uses it	Your grand-parents' gen uses it	Your children's gen uses it	Your grand-children's gen uses it
Cell phones (for talking)				X		
Smartphones (for talking)	X	X	X	X	X	
Smartphones (for taking and sending photos)	X	X	X	X	X	
Smartphones (for browsing Internet)	X	X	X		X	
Texting	X	X	X		X	
Skype	X	X		(sky what?)	X	
Voice mail	X	X	X	X	X	
Email	X	X	X	X	X	
The web	X	X	X	X	X	
Facebook	X	X	X		X	
Other social networking sites	X	X			X	
Professional networking sites	X	X				
Twitter	X	X			X	
Blogs	X	X			X	
E-cards	X	X				
Physical greeting cards	X	X	X	X		
Snail mail letters			X	X		
Snail mail postcards			X	X		
Printed digital photos			X	X		
Photos sent as digital files	X	X	X		X	

Now let's imagine that Lee's parents, Golda and Zamir, fill out their own Profiler. Golda is fifty-seven and Zamir is sixty-one.

GOLDA AND ZAMIR'S GENERATION PROFILER

DEVICE OR MEDIUM	You use it	Your gen uses it	Your parents' gen uses it	Your grand-parents' gen uses it	Your children's gen uses it	Your grand-children's gen uses it
Cell phones (for talking)	X	X				
Smartphones (for talking)	X	X			X	X
Smartphones (for taking and sending photos)	X	X			X	X
Smartphones (for browsing Internet)		X			X	X
Texting		X			X	X
Skype		X		(sky what?)	X	X
Voice mail	X	X	X		X	X
Email	X	X	X		X	X
The web	X	X			X	X
Facebook		X			X	X
Other social networking sites					X	X
Professional networking sites					X	X
Twitter					X	X
Blogs					X	X
E-cards					X	X
Physical greeting cards	X	X	X	X		
Snail mail letters	X	X	X	X		
Snail mail postcards	X	X	X	X		
Printed digital photos	X	X				
Photos sent as digital files	X	X				

From this, you can see that the more distant a generation is from your own, the more likely it is that they'll be comfortable with different devices. This means that if you want to stay connected to your grandparents, you may need to print out photos of your safari and stick them in a card that you mail.

Or if you want to stay connected to your grandkids, ask them to teach you to use Skype, which is a video telephone service that's available free to anyone with a computer. It's very easy to use.

Your Generation Profiler will let you know when you'll need to make an extra effort to be connected to someone you care about.

Of course, there will always be exceptions. You may have a grandpa who can text as fast as his arthritic thumbs will let him. Or you may be a Luddite, unwilling to use a cordless phone. That's okay.

Appendix B

YOUR PERSONAL LIMIT SETTER

You might want to use the chart below as a quick reference for setting boundaries. Find the type of relationship and device for which you would like to set a boundary. The chart will give you options for what to say or do. (Adapt them to your own situation, of course.)

Remember that people can react in very different ways when you set a boundary. Some will immediately respect it; others will defend their actions; still others won't get it. If someone won't respect a boundary, don't try to explain or defend yourself. Simply repeat the limit you want respected — and then hold that boundary firmly in place in the present and future. For details on how to restate a boundary and the other steps to take after you set one, refer to chapters 3 and 8.

Unless otherwise noted, the response can be delivered using the same medium or device involved in the situation. If a row in the chart is shaded, it means that you should handle that issue either in person or on the phone.

"INT" refers to an intrusion violation; "GAP" refers to a gap or distance violation

PERSON	DEVICE	VIOLATION	WHAT TO SAY OR DO
Relative	Email	INT Forwards: too many	*By email or phone call:* "Mom, I know you want to be thoughtful, but please stop sending pictures of mammoth Lego dinosaurs. I'm not actually interested in Legos anymore." *By voice mail:* "Uncle Ned, I know you love me, but please stop sending me email forwards. They are so interesting I get distracted from what I should be doing. Are you willing to help? Call me if you want to talk about this."
Friend	Email	INT Forwards: too many	*By email:* "Dear friends, Thanks for thinking of me, but please stop all forwarding. I just don't have time to even delete them. Help me out?"
Relative or friend	Email	INT Forwards: too many	*By email or phone:* "Dad, I know you love those pix of busty ladies, but stop sending them to me. I'm engaged now, and it seems disrespectful to my fiancée. (And they make me feel a bit sorry for Mom, too.)"
Acquaintance	Email	INT Forwards: too many	*By email:* "Steve, stop forwarding those explicit pictures. I don't want them. Please stop."
Boss (reasonable)	Email	INT Forwards: too many	*By email:* "Hey, Boss, got that last batch of forwards and not sure what you want me to do with them. I'm actually concentrating on the Gray project. Do you want me to stop focusing on that and deal with the forwards instead?"
Boss (CYA: covers himself by sending copies of everything)	Email	INT Forwards: too many	Create a file called Boss Forwards. Put all his forwards in that file. If he refers to something he's sent you, ask for the date or the subject line of that email. Then you can read that particular one.

PERSON	DEVICE	VIOLATION	WHAT TO SAY OR DO
Boss (disorganized)	Email	INT Forwards: too many	*If it's your job to organize her:* Quickly scan forwards and create files for each topic. Then file future forwards according to topic. *If it's not your job to organize him:* Talk to the person whose job it is. Ask if he gets copies of everything, too, or if he wants you to pass these forwards on to him. Find out if you need to check each forward, or if he will send back any that apply to your job.
Boss (sets you up or gives you work to make herself feel powerful)	Email	INT Forwards: too many	Scan every forward, then file it according to topic. Take notes on those she might expect you to remember.
Boss	Email	INT Forwards: sexist	*You are the same gender:* "Boss, you're taking a big chance. If one of the women gets a copy of this, you could be in big trouble for sexual harassment." *You are the opposite gender and he's a good boss:* "Boss, don't send me that stuff. I don't want to see you get in trouble." *You are the opposite gender and he's a bad boss:* Forward the emails to the human resources department, the legal department, and/or his boss.
Boss	Email	INT Forwards: racist or anti-religious	*You are the same race or religion:* "Boss, you're taking a big chance with this. If one of the other workers gets a copy of this, you could be in big trouble for racism. In any case, please leave me out of this. I don't share your perspective." *You are of the targeted race, and she's otherwise a good boss:* "Boss, you know this is racist, right? It is offensive to me and could cost you your job. Please stop sending this stuff." *You are of the targeted race or religion, and he's a bad boss:* Forward the material to human services, the legal deptartment, and/or his boss. *(chart continued on next page)*

PERSON	DEVICE	VIOLATION	WHAT TO SAY OR DO
Boss	**Email**	**INT** **Forwards: solicitations for charities, products, services, and/or causes**	***Reasonable boss:*** **"Hey Boss, got your solicitation but I already contribute to something else and that uses up my charity budget. You okay with that?"** ***Solicitation you agree with:*** **"Hi Boss, this is a little awkward. I actually already support this cause, and I don't want you to think I'm doing that to get on your good side. Contributed just last [whenever].** ***Power-hungry boss:*** **"Got your solicitation, sir. I have contributed." (Send $1.00, and document the solicitation in case you need evidence someday.)**
Boss	**Email**	**INT** **Forwards: invitation to political event**	***Reasonable boss:*** **"Hi Boss, got your invitation to the political rally. That's not a cause I can get behind, but I know how fair you are, so I'm hoping we can agree to disagree."** ***Napoleonic boss:*** **"Hi Boss, my church group has a meeting that night, and I'm a key pray-er. Sorry I won't be able to go. I'll pray for you and the candidate."**
Coworker	**Email**	**INT** **Forwards: too many**	**"Ned, I've gotten so I just don't have time to check forwards anymore. Please only send those that relate to our current project or that the boss has asked you to pass on."**
Coworker	**Email**	**INT** **Forwards: explicit, sexist, or racist**	**"Jake, you know you could get in trouble for this, right? Leave me out of it. Please don't send me any more."**
Coworker	**Email**	**Gossip**	**"Sal, I just want to focus on my work. I can hardly keep up as it is. Plus, you know it could be possible for management to track all our emails, right?"**
Coworker	**Email**	**INT** **Forward: shoe sale at Nordstrom or other sales event**	**"You know my weakness, girl, but help me here. No forwards please. (Call me when you plan to go and I'll see if I can join you.)"**
Someone you care about	**Email**	**GAP** **Lack of response to invitation**	**"Hi buddy. Did you get my invitation? Call me today and let me know if you're joining me."**

PERSON	DEVICE	VIOLATION	WHAT TO SAY OR DO
Someone you care about	Email	GAP Lack of response to questions	"Hi love, can't proceed till you answer my questions. If you didn't get that email, either let me know by email today or phone tomorrow."
Someone you care about	Email	GAP Lack of response to important message	*Phone:* "Hi sweetie, did you get my message about the heart transplant? Need a response today."
Someone you care about	Email	GAP Lack of response to idea	"Jack, if you disagree with my idea, I'm okay with that, just need to know. Please respond."
Someone you care about	Email okay for setting up a time to talk in person or on the phone.	GAP Routinely doesn't respond	"Hi. What does it mean that you haven't been replying to my messages? Too busy? I'm writing more often than you want me to? Something has changed for you? Don't like email? I would like to talk about this. Are you willing to set a time? I'm available all weekend, Thursday afternoon, and every evening except Mondays."
Boss	Email	GAP Lack of response	"Hi Boss, as I said on 9/5, I need to know the length of the swatch before I can proceed. Let me know if I should be asking someone else."
Boss	Email	GAP Still not responding	"Hi Boss, as I said on 9/6, I need to know the length of the swatch before I can proceed. I'm going to check with the VP to see if he knows. Call me if you want to give input."
Boss (wishy-washy)	Email	GAP Lack of response	"Hey Boss, I start cutting pieces tomorrow. Attached is the current pattern. If I don't hear from you by 7:00 am, I'll proceed with that plan."
Someone in a supervisory or advisory position — mentor, spiritual adviser, coach, boss, therapist, teacher	Any device	Sends sexually suggestive material, comments, or messages	*By email:* "Sir, please stop sending me such messages. You know it's not right. Stop now." Do not erase that person's message. Forward it to the person's superior. *(chart continued on next page)*

PERSON	DEVICE	VIOLATION	WHAT TO SAY OR DO
Anyone	Email or text	INT Sends explicit picture of their own private body part	Unless this is someone with whom you already have an intimate relationship, and this is a part of your sex play, be wary. The more inappropriate this is — in terms of the nature of your relationship — the more unbalanced that person may be. Back out of the relationship. If this is a student or anyone you are in charge of as a teacher, mentor, counselor, or adviser, set a non-negotiable boundary. If it is violated again, end the relationship. *Respond initially with an email like this:* "Len, it was not okay that you sent me that picture. Do not do such a thing again. If you do, I can no longer be your [teacher, adviser, mentor, etc.], and I will have to transfer you to someone else." If this is someone who supervises or advises you, do not keep it a secret. Tell their superior. Don't be alone with the person who sent you the picture. In this case, you won't have to set the boundary; let someone above them take care of it.
Someone you care about	Text	Mismatch: she texts; you talk	"You're a texter; I'm a talker. Text all you want; I'll call you with my responses."
Someone you care about	Text	Mismatch: she texts; you email	"You're a texter; I'm an emailer. Text all you want; I'll email my responses."
Someone you care about	Text	INT Too many	"At work now. Can't read or respond. Please wait till 5:00 pm to text me again."
Someone you care about	Text	You both usually do running check-ins throughout the day.	"Just so you know. Getting ready to drive. Won't be reading texts till I arrive."
Please note: Texting is not a good way to have a detailed dialogue or work out an issue. In the following situations, send a text that sets up a phone or face-to-face conversation.			
Someone you care about	Text	Wrong device; bringing up issue or problem	"Jen, I agree this is important, but text is too limiting. Need room for more words and also want to hear your voice. Let's do this on the phone."

PERSON	DEVICE	VIOLATION	WHAT TO SAY OR DO
Someone you care about	Text	INT Too many on an ongoing basis	"Have something to talk about with you. Will call later." Setting this boundary requires the give-and-take of a discussion, either over the phone or in person. Otherwise, there's too great a chance for misunderstanding or hurt feelings. *The conversation might go like this:* "Mar, on one hand, I love knowing what's going on with you each day. On the other hand, that many texts is interrupting my focus on my own life. Would you be willing to limit your texts to one a day?"
Someone you care about	Text	GAP Emergency	"Hal, need more info. Am calling now. Answer."
Someone you care about	Text	GAP Not answering	"Are you getting my texts? Sent 3 this week. Let's talk."
Someone you care about	Text	GAP Perpetually not responding	"Gil, what does it mean when you don't respond to my texts? Let's talk about it. I'll call soon."
Boss	Text	INT Too many, too chummy, or borderline sexually suggestive	"Hi Boss, thanks for your interest. Am totally focused on work. Should we talk? How about tomorrow in the lunchroom?" (Important to do this in a public place.) *Then, in person, you can say:* "I'm sure you don't mean anything by it, but the tone of your texts could be misunderstood. Would you be willing to just stick to business?" Save the texts in case you need to document sexual harassment.
Boss	Text	GAP Wrong device; routinely texts too cryptically for you to decipher instructions, leaving you hanging	*By text:* "Boss, got text. Need more info to proceed. Please email or call with more details. Info needed: Size of widget Order deadline Range of tolerance" *Or by email:* "Boss, got your text. Need more info to proceed. Am emailing you a list of info needed." (The advantage to this is it'll be easier for her to email back with the responses you need.)
Coworker	Text	INT Snide comment about boss	"I know you're frustrated, but leave me out of texts about the boss. Too dangerous." *(chart continued on next page)*

PERSON	DEVICE	VIOLATION	WHAT TO SAY OR DO
Coworker	Text	INT Sexist comments about some people in the office	"Ed, stop texting me about our female coworkers. I don't want to participate."
Coworker	Text	INT Interruptive comments about non-work things	"Sue, focusing here. Just don't have time for texts about Little League. Please stop."
Anyone	Text or Email	INT Forwarded your message, meant to be private, to someone else	"Jon, you available? Need to talk by phone." *Phone:* "Jon, that message I sent you was private, just between you and me. I'm [upset, angry, concerned, frustrated] that you forwarded it on to Mom. You know how she worries. Now she's upset. Are you willing to not forward any more of my messages?" If Jon is a Leaky Lucy, he won't be able to hold this boundary. You'll have to stop sending him info you don't want passed on.
Someone you care about	Any device	INT & GAP Using device during a time set aside for you to connect with each other	"This time is just for us. I want to disconnect from the world and just be together. Are you willing?"
A group	Any device	INT & GAP Using device during a time set aside for you to be doing something with each other	"We're here because we like to ________________ together. I want to disconnect from the world, turn off the devices, and just be with y'all. Would you all join me? Would you be willing to turn off phones, etc., while we're together?"
Anyone	Twitter	INT Tweets something mean about you (whether it's true or untrue)	*If it's a stranger, let it go. If it's a friend, acquaintance, or relative, tweet back:* "What's up? Why are you saying this? Let's talk on the phone." *By phone:* "Please, stop saying mean things about me. You know all tweets are public, right? If you have an issue with me, please come to me directly."

PERSON	DEVICE	VIOLATION	WHAT TO SAY OR DO
Anyone	Facebook	INT Puts your private info on their wall	If this is someone you don't know, or who is way outside your circle, they may not have an investment in respecting your boundary. But you can try this: "Hi. We barely know each other, but I noticed you posted stuff about me on your wall. If you don't mind telling me, I'd like to know where you got it. I'd also much appreciate your deleting it." If this is someone with whom you have an ongoing relationship, set up a face-to-face or phone conversation: "Nell, you posted my private info on your wall. I want you to delete it. Are you willing?" Once you have an agreement from her to delete it, follow up and watch for her to do it. If she does, call back and thank her, and then repeat the limit: "Please don't post my private info again. If you aren't sure what's private, please check with me. Thanks." If she doesn't take care of it, confront her again. If she continues to disregard your request, you can flood her own wall with posts — comments, likes, pictures, etc. — to push that piece down the page. Do not retaliate. Do not post private information about her. This will only escalate. You have now learned that you can't trust her with private information. *(chart continued on next page)*

PERSON	DEVICE	VIOLATION	WHAT TO SAY OR DO
Closed private group	**Chat room**	**INT** **Members are conveying private chat info outside group in some way (Facebook, Twitter, etc)**	**"Hey, guys, I want to discuss a boundary for our group. I'd like us to agree to keep what we say here private. I don't want to read my private, personal thoughts that I share with you on your Facebook wall. Anybody want to join me on this?"** **If the group agrees, then say:** **"Then, Tigger, would you be willing to remove that post on your wall that repeated my comments from last week?"** **If the group has already agreed and someone has violated the boundary:** **"Cal, despite our agreement, you posted some of my remarks on your wall. I want to know if you understand the boundary.** **"If you do, then you know it was not okay to post my comments.** **"So, will you take them off, please? And can we trust that you will honor the boundary in the future, or not?"** **If someone isn't capable of respecting the boundary, as a group you'll have to decide on the consequence: Kick that person out? Leave the group yourself? Stop sharing anything you don't want revealed?**

Appendix C

RESOURCES FOR DEVELOPING INTIMACY SKILLS

In chapter 17, I spoke of my concern that younger generations — accustomed to remote friendships, communicating primarily with sound bites through Facebook, Twitter, and texting — might not master the sequential tasks that lead to true intimacy. This would set them up to enter marriage lacking the skill set to navigate the hard work of daily relationship that leads to truly amazing union.

We don't know what we don't know. Until a world is opened up to us, we can't know what we're missing. Growing up in a culture that is heavily involved in electronic and digital communication, you may not be aware of the skills you don't possess. The missing pieces will show up as consequences or negative outcomes that happen again and again. One way to catch on to the skills you need is to be with people you care about — phone off, no device or app use — and then notice how the interaction proceeds, paying attention to how you feel, how the other person reacts, and whether it is satisfying or dissatisfying.

The following checklist may help you begin to spot your patterns. You can use it in a variety of ways:

- Before an encounter, to increase your observational skills
- After an encounter, to take note of the feelings you experienced or to review the flow of the event
- Over a couple of weeks, to identify problems that persist

Once you've identified some of your patterns, take your list to a counselor, parent, or teacher you trust, and talk about it. If you can't pinpoint any patterns but your encounters are not satisfying, talk to your mentor about that.

EXERCISE: Observations While Being with Someone I Care about

- ❐ I ran out of things to say, was not sure what to bring up.
- ❐ I noticed that all our talk was about apps or gaming.
- ❐ I felt a pull to get out my tablet. [What happened right before you felt that way?]
- ❐ I suddenly felt like backing away; something put distance between me and my friend. [What did that? What did they say or not say, do or not do?]
- ❐ I suddenly felt closer to my friend. [Did that feel good? Was it also a little scary? Do you know what scared you about it?]
- ❐ My friend suddenly seemed more distant from me. [What happened right before that?]
- ❐ I had a different opinion and didn't know whether to say it.
- ❐ I voiced my different opinion and it didn't go well.

When you are with two people who seem to be very close — an aunt and uncle, your parents, married friends — watch how they handle things. How do they manage differences or disagreements? What evidence do you see of their closeness? Can you see them supporting each other? What is their style of communication? Do they each listen well to the other?

You can try out the behaviors you see them using, and notice what happens as you apply your efforts to your friendships.

Watch some older TV shows that focus on friendships, especially those that occurred before the digital age. Notice the skills the friends use, and how those skills work. A great example is *Sex and the City*. There you can see friends with very different personalities listening, supporting each other, involved with life yet making time for their relationships, disagreeing, letting each other down, getting angry, and fixing mistakes.

Intimacy Skills Fast Track

If you want to get good fast, your best course is group therapy. Such venues are not just about therapy; they are about learning boundaries, communication skills, how to say hard things, and how to observe yourself as you interact. You will gain a wealth of benefits.

Group therapy is quite different from individual therapy. Sometimes a person will need to work individually in order to get to the point where group work will be beneficial. If this is the case for you, don't stop there. Move into a group therapeutic situation as soon as you feel ready.

Some group therapists are better than others, so read all you can about the various group opportunities in your area and check the therapists' websites. You can also ask others — friends, doctor, spiritual adviser, teacher, school counselor — if they know of a good group therapist.

If you are anywhere near a Systems-Centered Training® (SCT)

workshop, therapist, or conference, you might want to check it out. This body of professionals is unparalleled in their brilliance with group process, streamlined methods, and skill delivery that can take years off an ordinary therapy experience, whether it be group or individual. SCT is a methodology that, within a group context, creates an environment where your skills can expand rapidly, and your personal development as well.

It can be a little rough at first, because the group will probably be quite unlike any expectations you have for it, particularly if you've seen cinematic group sessions and think you know what group therapy looks like. Stick it out. Give it a chance. The fruits will blossom in marvelous and unexpected ways.

SCT professionals are so good that it's worth driving farther than you want to and paying more than you would for a local therapist. You'll save in the long run, in both dollars and minutes.

For more information, go to www.systemscentered.com.

ENDNOTES

Page 66 *37 percent of your brain is now unavailable for driving*: Marcela Adam Just, Timothy A. Keller, and Jacquelyn Cynkar, "A Decrease in Brain Activation Associated with Driving When Listening to Someone Speak," *Brain Research* 1205 (2008): 70.

Page 68 *70 to 84 percent of people answer the phone while driving*: US Department of Transportation, "National Distracted Driving Telephone Survey Finds Most Drivers Answer the Call, Hold the Phone, and Continue to Drive," *Traffic Tech*, Report 407, December 2011, 2.

Page 68 *drunk drivers and drivers using cell phones are equally impaired*: David L. Strayer, Frank A. Drews, and Dennis J. Crouch, "A Comparison of the Cell Phone Driver and the Drunk Driver," *Human Factors*, University of Utah, Salt Lake City, Summer 2006, 388.

Page 68 *There's a 37 percent decrease in brain availability*: Just, Keller, and Cynkar, "A Decrease in Brain Activation," 70.

Page 69 *"Put simply, tasks that draw drivers' eyes away..."*: Rebecca L. Olson, Richard J. Hanowski, Jeffrey S. Hickman, and Joseph Bocanegra, "Driver Distraction in Commercial Vehicle Operations," US Department of Transportation, Report FMCSA-RRR-09-042, September 2009, xxiv.

Page 69 *The death rate for drivers between the ages of twenty-five and forty*: Lawrence Lam, "Distractions and the Risk of Car Crash Injury: The Effects of Drivers' Age," *Journal of Safety Research* 33 (2002): 415.

Page 82 *You're deep into your research about the tidal habits*: Note that *kelp creeper* is my invented name for a kelp inhabitant, not a true scientific or commonly used term (and there is no such research).

Page 86 *Eyestrain, back and joint issues, carpal tunnel syndrome, and e-thrombosis*: For more on e-thrombosis, see www.take-time-out.info/e-thrombosis.html.

Page 90 *93 percent of American children between twelve and seventeen*: Pew Research Center, "Teen and Young Adult Internet Use," Pew Research Center, undated, www.pewresearch.org/millennials/teen-internet-use-graphic. Also of interest is Amanda Lenhart, Kristen Purcell, Aaron Smith, and Kathryn Zickuhr, "Social Media and Young Adults," Pew Research Center, February 3, 2010, www.pewinternet.org/Reports/2010/Social-Media-and-Young-Adults.aspx.

Page 90 *Approximately one in five children were sexually solicited*: David Finkelhor, Kimberly J. Mitchell, and Janis Wolak, "Online Victimization: A Report on the Nation's Youth," National Center for Missing and Exploited Children, June 2000, ix.

Page 90 *While this percentage decreased to one in seven in a second study*: Janis Wolak, Kimberly Mitchell, and David Finkelhor, "Online Victimization of Youth: Five Years Later," National Center for Missing and Exploited Children, 2006, 1, www.missingkids.com/en_US/publications/NC167.pdf.

Page 90 *75 percent of solicitors asked to meet the youth in person*: Ibid., 18.

Page 90 *79 percent of these solicitation attempts happened*: Ibid.

Page 91 *One in three youths are exposed to unwanted sexual material*: Ibid., 8.

Page 91 *4 percent of youths received distressing sexual solicitations*: Ibid.

Page 91 *9 percent of youth Internet users were very or extremely upset*: Ibid., 9.

Page 91 *90 percent of sexual solicitations and approaches*: Ibid., 16.

Page 91 *90,000 sex offenders had MySpace pages*: Marlon Walker, "MySpace Removes 90,000 Sex Offenders," *NBC News*, February 3, 2009, www.nbcnews.com/id/28999365/ns/technology_and_science-security/t/myspace-removes-sex-offenders/#.UZoqKL-HolI.

Page 91 *"The increase in exposure to unwanted sexual material..."*: Wolak, Mitchell, and Finkelhor, 1.

Page 91 *While the percentage of stranger solicitations has dropped*: Ibid., 17.

Page 92 *73 percent of incidents were not reported*: Ibid., 33.

Page 92 *only one in five hundred incidents were reported*: Ibid., 34.

Page 92 *almost half of today's youth using the Internet*: Ibid., 9.

Page 94 *about half of the interviewed kids reported being* told: Ibid., 45.

Page 101 *Girl Scout Internet Safety Pledge*: "Girl Scout Internet Safety Pledge for All Girl Scouts," Girl Scouts of America, undated, www.girlscouts.org /help/internet_safety_pledge.asp.

Page 103 *"Exposure to violence in media..."*: Committee on Public Education, "Media Violence," *Pediatrics* 108 (2001): 1495, http://pediatrics.aappublications .org/content/124/5/1495.full.

Page 103 *A National Television Violence study evaluated*: Joel Federman, ed., *National Television Violence Study*, vol. 3 (Thousand Oaks, CA: Sage, 1998).

Page 103 *"The highest proportion of violence..."*: Committee on Public Education, "Media Violence," *Pediatrics* 108 (2001), http://pediatrics.aappublications. org/content/124/5/1495.full#ref-22.

Page 103 *"showed at least one character..."*: Fumi Yokota and Kimberly M. Thompson, "Violence in G-Rated Animated Films," *Journal of the American Medical Association* 283 (2000): 2717, www.bvsde.paho.org/bvsacd/cd42/animated.pdf.

Page 104 *"More than 80 percent of the violence portrayed..."*: Committee on Public Education, "Media Violence," *Pediatrics* 108 (2001), http://pediatrics .aappublications.org/content/124/5/1495.full#ref-22.

Page 106 *"Children are influenced by media..."*: Committee on Public Education, "Media Violence," 1223.

Page 106 *Considering that kids spend forty-five hours a week*: "Violence in Media Really Matters," Common Sense Media, 2010, www.commonsensemedia.org /videos/violence-media-really-matters.

Page 107 *Ninety-seven percent of children now play them*: Amanda Lenhart, Joseph Kahne, Ellen Middaugh, Alexandra Macgill, Chris Evans, and Jessica Vitak, "Teens, Video Games, and Civics," Pew Research Center, September 16, 2008, www.pewinternet.org/Reports/2008/Teens-Video-Games -and-Civics.aspx.

Page 107 *Such gaming causes*: Craig Anderson, "The Impact of Interactive Violence on Children," Hearing before the Senate Committee on Commerce, Science, and Transportation, March 21, 2000, 36, www.gpo.gov/fdsys/pkg /CHRG-106shrg78656/pdf/CHRG-106shrg78656.pdf.

Page 108 *"Reason 1. Identification with the aggressor..."*: Ibid., 37.

Page 108 *Repeated exposure to violent behavioral scripts*: Committee on Public Education, "Media Violence," 1223.

Page 108 *video games have been found to be addictive*: M. D. Griffiths and N. Hunt, "Dependence on Computer Games," *Psychological Report* 82 (1998).

Page 108 *Interpersonal violence, as victim or as perpetrator*: Committee on Public Education, "Media Violence," 1223–24.

Page 109 *"Playing violent video games has been found..."*: Ibid., 1223.

Page 109 *The more realistic the portrayal of violence*: R. M. Liebert and J. Sprafkin, *The Early Window: Effects of Television on Children and Youth*, 3rd ed. (New York: Pergamon Press, 1988). In "Media Violence," p. 1223.

Page 111 *"'Peter,' his mother sighed, 'how could you..."*: Jodi Picoult, *Nineteen Minutes* (New York: Washington Square Press, 2007), 69.

Page 113 *Dr. Latanya Sweeney, a computer scientist at Harvard*: Jonathon Shaw, "Exposed," *Harvard* magazine, September–October 2009, 39, harvard magazine.com/2009/09/privacy-erosion-in-internet-era.

Page 115 *"Many of the most popular applications..."*: Emily Steel and Geoffrey Fowler, "Facebook in Privacy Breech," *Wall Street Journal*, October 18, 2010, online.wsj.com/article/SB10001424052702304772804575558484075236968.html.

Page 116 *When you press the Like or Tweet button*: Jennifer Valentino-Devries and Jeremy Singer-Vine, "They Know What You're Shopping For," *Wall Street Journal*, December 7, 2012, online.wsj.com/article/SB10001424127887324784404578143144132736214.html.

Page 130 *The* Wall Street Journal *analyzed the tracking files*: "What They Know," *Wall Street Journal*, Blogs, April 12, 2012, blogs.wsj.com/wtk or online.wsj.com/public/page/what-they-know-digital-privacy.html.

Page 132 *"The* Journal *found that Microsoft Corp.'s..."*: Julia Angwin, "The Web's New Goldmine: Your Secrets," *Wall Street Journal*, July 30, 2010, online.wsj.com/article/SB10001424052748703940904575395073512989404.html.

Page 133 *"He sent a note to a showroom near Atlanta..."*: Valentino-Devries and Singer-Vine, "They Know What You're Shopping For."

Page 133 *"The use of real identities across the Web..."*: Ibid.

Page 134 *Current estimates are that 88 to 90 percent*: Messaging Anti-abuse Working Group, "Email Metrics Program, The Network Operators' Perspective," November 2011, www.maawg.org/sites/maawg/files/news/MAAWG_2011_Q1Q2Q3_Metrics_Report_15.pdf.

Page 135 *(Google has already done this.)*: Julia Angwin and Jennifer Valentino-Devries, "Google's iPhone Tracking," *Wall Street Journal*, February 27, 2012, online.wsj.com/article/SB10001424052970204880404577225380456599176.html.

Page 161 *For example, the match.com privacy policy states:* "Match.com, L.L.C. Privacy Policy," undated, www.match.com/registration/privacystatement.aspx?lid=4.

Page 162 *"You agree that we may use..."*: "Privacy Statement," undated, www.spark.com/Applications/Article/ArticleView.aspx?CategoryID=1948&ArticleID=6498&HideNav=True#privacy.

Page 198 *Person vs. Member*: My first exposure to the person system vs. member system occurred through my training at SCT with Yvonne Agazarian and other licensed trainers. The information here has been derived from that training and my resultant experiences with it.

GLOSSARY

ALGORITHM — A set of steps that will lead to a specific outcome that solves a problem. A recipe for chocolate chip cookies or the requirements for receiving a college degree could be considered an algorithm, but the word usually refers either to a mathematical process that leads to a numerical answer that may then translate to a human situation or to programming a computer. The ideal is to achieve the result in the fewest and simplest steps.

APP — A shortened form, or nickname, for *application*. In the computer world, an application is the software that makes possible any way you use a computer. Microsoft Word, Trade Nations, Photoshop, and the system software that runs your computer are all useful applications.

AUTOCORRECT — In a digital device or smartphone, a function that detects a misspelled or incorrectly written word and either automatically replaces it with the most probable word or offers a drop-down list for you to choose from. The term is also used to refer to autofill, another function that completes the word you've started so you don't have to type all the letters. This is most convenient with phones because they have tiny buttons, but if you don't take a look at what you've typed, you may send a message that has no relationship to the one you intended to send.

BETA STATE — A brain-wave pattern associated with alertness and cognitive thinking, greater than 13 Hertz (Hz; cycles per second).

BLOG — Initially, a *web log*, a web page that was one's personal journal. Currently, blogs can be written by individuals or groups, and contain any manner of content, including professional comment, instruction, or discussion. Many are now interactive, so that site visitors can comment or even message each other, becoming another venue for social networking.

BOT — Ro*bot*; a computer program that operates automatically, typically to carry out automated tasks. For example, if you submitted an article to Wikipedia, a bot would check that your article contained no copyright violations, putting one more good editor out of work.

BOUNDARY VIOLATION — an action (or failure to act) that weakens or penetrates a boundary, harming the entity that was protected by it.

CHAT ROOM — A "room" in cyberspace in which people can mentally gather to communicate. Most chat rooms actually operate through typed conversations, although it is possible to actually chat with some computers in some chat rooms. Some chat rooms have moderators; many don't. A chat room can be public — open to anyone — or private, available only to a select group.

CLOUD — Just a bunch of servers that get referred to metaphorically as the cloud.

COOKIE — A packet of data that transfers information between computers. For example, a cookie may carry an identifier that links to your name, password, and shopping cart items so that when you return to a particular online store, that information fills in automatically. A cookie might also be used to record (for the website owner) the pages you visited at a previous time and any buttons you've ever pushed on that site.

CRAIG'S LIST — Online classifieds through which you can buy or sell nearly anything — a used wig ($15); a Jesus costume, including the crown of thorns ($10); sperm; or a shopping mall. Craig's List looks like the classified ads in a hometown paper and is free.

CRAWLER — A robot used by a search engine to index content on the Internet (also known as a spider). For example, when you enter "hairpin lace" into a Search box on a search engine such as Google or Bing, you can find anything you want to know about that subject, including pictures, how-to videos, and where to buy it.

CYBERWORLD — A term I coined to refer to the now-massive penetration of both computers and the Internet into the previously computerless world. Computers started out as convenient, efficient business and educational devices and have now ushered in a new world paradigm.

Distance violation — The disruption of a boundary caused by too much distance, in a context where a degree of interaction is essential to the health of the entity.

eBay — Online catalog through which you can buy or sell nearly everything: a pope's crucifix, Rosa Parks's handwritten notes, an island. Ebay has a slick professional magazine look and charges a fee.

Facebook — A social media and social networking site at which individuals, groups, businesses, and other entities can, essentially, have an interactive web page on which they can blog, broadcast their opinions, post pictures, connect with friends, promote their business, announce special events, and waste a lot of time. Facebook also offers apps and games so that one can connect with strangers around mutual interests.

Firefox — A web browser (search engine) that lets you find, retrieve, and use web pages on the Internet.

Firewall — The first line of defense to prevent unauthorized access to a computer or network. Both hardware and software can provide firewall protection.

Forward — An email that you receive and then pass on to others.

Friend — On Facebook, someone you have selected as a contact, either as a result of inviting them or of being invited to their Friend list. You may not actually know people on your Friends list.

Gap violation — The disruption of a boundary caused by too much distance, in a context where a degree of interaction is essential to the health of the entity.

Google (verb) — The process of using a search engine, such as Google or Bing, to find specific content on the Internet.

Instant message (IM) — A text-based (as opposed to voice-based) communication, possible on many digital devices.

Internet Explorer — A web browser (search-engine) developed by Microsoft that lets you find, retrieve, and use web pages on the Internet.

Intrusion violation — Penetration of a protective boundary.

Isomorphic — A term referring to similarities, in a hierarchical system, in how the boundaries function across the levels of the hierarchy.

Like — On Facebook, an option for approving or acknowledging — by pressing a Facebook button — any business or entity, or an item posted on someone else's wall or page.

LinkedIn — A professional, business-related networking site for the purpose of connecting colleagues, clients, and potential employers.

Malware — *Mal*icious soft*ware* that can be planted in your computer to make it do bad things to benefit someone other than you.

MySpace — A web-based social networking site that is currently focusing on music and entertainment.

News Feed — Your own Facebook page on which are posted all the materials your Friends have posted on their Walls or Timeline. If you Like or Share their posts, these are then posted on all your Friends' News Feeds.

Page — Used on the Internet to refer to a web page, the blank screen on which all content appears.

Picture Recognition — Software that compares aspects of faces on web pages in order to name or identify the person.

Post — To publish a message.

Safari — A web browser that lets you find, retrieve, and use web pages on the Internet.

Search engine — A program that searches documents on the World Wide Web (www) for specific keywords, then lists links to all the pages where those keywords are found.

Snail mail — A message, usually written on paper and put into a container like an envelope or a box, that is transferred physically through a postal service.

Social media — Social networking sites, such as Facebook and Twitter.

Spam — Junk mail sent to numerous recipients via email.

Spider — See Crawler

Spybot — A *spy*ing ro*bot* that looks for private information or your computer's unique identifying number to service malware. It usually operates through worms. Currently, malware terms are not standardized and thus have a variety of meanings.

Tablet — a mobile computer, such as an iPad, that usually operates by touch screen.

Texts — Brief electronic messages sent between two or more mobile devices over a phone network.

Touch screen — A screen on a computer, tablet, or phone that can be operated by direct touch.

Tweet — A single Twitter message.

Twigged — My own term for receiving a tweet, based on the bird metaphor.

Twitter — A brief social messaging tool (limited to 140 characters) that asks, "What are you doing?" The answers go beyond what you might actually be doing at the moment to comments, opinions, rants, and raves.

URL — Uniform Resource Locator (web address).

Wall — The metaphorical wall upon which early Facebook users posted their comments and responses. Currently, the News Feed and Timeline pages contain what used to be on the Wall.

Web beacon — A very small (usually 1 pixel) transparent graphic image that, with a cookie, monitors a user's behavior on a website or in response to an email.

WEB CRAWLER — See Crawler

WI-FI — Wireless local networking technology.

WORM — Now considered malware, computer worms were once thought to be useful for research or to fix computer problems, because they can spread throughout systems beyond the original computer or system they were planted in. However, bad worms can collect private information, disseminate viruses, attach to email, send spam, delete files, or contain documents that open a (back)door in your computer to let in other malware.

INDEX

ABOUT THE AUTHOR

Photo by Mary Grace Long

For over forty years, Anne Katherine shepherded clients into their inner spaces that held the forgotten wounds and hidden sorrows that shape a person's perceptions and choices. Once there, she knew how to bathe those sacred places with compassion in a way that brought both soothing and startling insight, opening them to new dimensions within themselves and new possibilities for their lives.

Anne discovered that gaining skill in setting boundaries provides essential protection for any improved paths a person wants to walk. Thus, as with most of her books, the trilogy on boundaries arose from her desire to give people tools to better their lives.

In her professional work, Anne has created numerous programs and processes, including:

- how to fix a broken appetite switch
- how to recover from addiction to misery and self-sabotage
- how to discover your life's purposes

Although she has retired from private practice and offers very few workshops, you can still find much of Anne's work in her nine books, which each focus on some aspect of self-healing, or on her websites.

Anne has been on television and radio, including Lifetime, CNN, NPR, and many other networks. Her credentials included Licensed Mental Health Counselor, Board Certified Regression Therapist, Certified Eating Disorders Specialist, Registered Hypnotherapist, and Master of Arts, Psychology.

Master Your Appetite is her online,
self-directed program for appetite switch repair.
www.masteryourappetite.com

Find news about her books,
and announcements of her rare appearances, at
www.annekatherine.org

NEW WORLD LIBRARY is dedicated to publishing books and other media that inspire and challenge us to improve the quality of our lives and the world.

We are a socially and environmentally aware company, and we strive to embody the ideals presented in our publications. We recognize that we have an ethical responsibility to our customers, our staff members, and our planet.

We serve our customers by creating the finest publications possible on personal growth, creativity, spirituality, wellness, and other areas of emerging importance. We serve New World Library employees with generous benefits, significant profit sharing, and constant encouragement to pursue their most expansive dreams.

As a member of the Green Press Initiative, we print an increasing number of books with soy-based ink on 100 percent postconsumer-waste recycled paper. Also, we power our offices with solar energy and contribute to nonprofit organizations working to make the world a better place for us all.

Our products are available
in bookstores everywhere.
For our catalog, please contact:

New World Library
14 Pamaron Way
Novato, California 94949

Phone: 415-884-2100 or 800-972-6657
Catalog requests: Ext. 50
Orders: Ext. 52
Fax: 415-884-2199
Email: escort@newworldlibrary.com

To subscribe to our electronic newsletter, visit:
www.newworldlibrary.com

3,147
trees were saved

HELPING TO PRESERVE OUR ENVIRONMENT

New World Library uses 100% postconsumer-waste recycled paper for our books whenever possible, even if it costs more. During 2011 this choice saved the following precious resources:

ENERGY	WASTEWATER	GREENHOUSE GASES	SOLID WASTE
22 MILLION BTU	600,000 GAL.	770,000 LB.	225,000 LB.

www.newworldlibrary.com

Environmental impact estimates were made using the Environmental Defense Fund Paper Calculator @ www.papercalculator.org.